P9-AFH-762

Concepts in Toxicology

Concepts in Toxicology

John H. Duffus
The Edinburgh Centre for Toxicology, Edinburgh, UK

Douglas M. Templeton
University of Toronto, Toronto, Canada

Monica Nordberg
Karolinska Institutet, Stockholm, Sweden

RSC Publishing

ISBN: 978-0-85404-157-2

A catalogue record for this book is available from the British Library

Published by The Royal Society of Chemistry,
Thomas Graham House, Science Park, Milton Road,
Cambridge CB4 0WF, UK

Registered Charity Number 207890

For further information see our web site at www.rsc.org

Preface

Toxicology as an applied science has developed rapidly over recent years. Partly this is due to introduction of comprehensive legislation to ensure that chemicals are produced, used and disposed of safely. The legislation has been driven by public concern about the possible hazards of exposure to new substances. Because of Rachel Carson's 1962 book, *Silent Spring*, concern initially focused on pesticides but was extended to other substances as it became clear that many diseases, especially cancer, might be caused by chemical exposures. Thus, legislation was introduced requiring toxicity assessment of substances, *e.g.* the U.S. Toxic Substances Control Act, and subsequently risk assessment of processes involving substances of concern, including their transport and final disposal as waste. The most far-reaching legislation to date is that recently introduced by the European Union. The REACH (Registration, Evaluation, Authorisation and Restriction of Chemicals) Regulation applies to all substances imported into or produced by the European Union countries (apart from certain special groups such as medicines and pharmaceuticals which have separate regulation). It makes registration and specified toxicity testing compulsory for all chemicals available in the market place. This is followed by evaluation of inherent hazard by the European Regulatory body, the European Chemicals Agency (ECHA), followed by assessment of life cycle risk associated with proposed uses. If it is considered that the given substance can be used safely for the purposes intended then the substance will be listed as 'authorized' for these purposes. Unauthorized substances cannot be used any longer within the European Union and cannot be imported or marketed.

In addition to the ever-growing body of legislation throughout the world, a program led by the United Nations Economic Commission for Europe (UNECE) aims to establish a Globally Harmonized System (GHS) for regulating the production, trade and use of chemicals. For this to be possible, it is essential that there should be a consensus regarding the meaning of the relevant terminology. This goes further than simple dictionary definitions because terms

Concepts in Toxicology
By John H Duffus, Douglas M Templeton and Monica Nordberg

Published by the Royal Society of Chemistry, www.rsc.org

acquire significance from usage. Usage changes subtly over the years. Terms may be given implications that are understood by practitioners in toxicology but not by those who are now legally responsible for interpreting toxicological information for risk assessment and risk management. Thus, the authors expect that the present book will be helpful to chemists, engineers, pharmacologists, toxicologists, risk assessors, regulators, medical practitioners and regulatory authorities. It should also have a wider interest for every person who needs to use potentially toxic substances without undue risk of getting a chemically-induced illness.

This book is, so far as the authors know, the first attempt to explain, in depth, concepts in toxicology related to risk assessment of manufacture, transport, use and disposal of chemicals. It is assumed that readers will have some relevant background in science, at least to higher school level. The terms have been chosen because of their frequent use in the literature, because they have commonly been subject to misunderstanding and because they still have the potential to cause problems for newcomers to toxicology and those responsible for safe chemical management. For readers who are not toxicologists and who may wish to learn more of the subject we recommend the sister book, *Fundamental Toxicology*, 2nd edn, edited by J. H. Duffus and H. G. J. Worth, published by the Royal Society of Chemistry in 2006.

Contents

Concepts in Toxicology
By John H Duffus, Douglas M Templeton and Monica Nordberg

Published by the Royal Society of Chemistry, www.rsc.org

Acknowledgments

We are grateful to all those who have contributed to this book throughout its development from the original papers published in *Pure and Applied Chemistry* to the present publication. Before submission for publication, most of the text was reviewed in part or in whole by Karin Broberg, Robert B. Bucat, Rita Cornelis, John Fowler, Sean Hays, Birger Heinzow, Hartwig Muhle, Stuart J. Nelson, Monika Nendza, Mike Schwenck, Ronald C. Shank, Wayne Temple, Howard G. J. Worth and other anonymous referees for *Pure and Applied Chemistry*.

All our reviewers kindly provided us with constructive criticism that has helped to ensure the quality of the text.

We are grateful to IUPAC for supporting the many terminology projects that have made this book possible and especially the preparation of the explanatory dictionaries from which this book is derived.

Introduction

Objective of the Book. The objective of this book is to give full explanations of the meaning and usage of key toxicological terms. This requires a description of the underlying concepts, going well beyond a normal dictionary definition, making plain underlying assumptions and implications, especially for regulatory toxicology, which has come to influence so much of our life where the use of chemicals is concerned. There are many reasons for this. Firstly, with the advent of antibiotics and the development of increasingly effective means of controlling infectious disease, attention was turned to diseases resulting from exposure to chemicals. Prevention of these diseases required regulation of exposure and so new laws, such as the Toxic Substances Control Act in the USA, were introduced. These laws required assessment of toxicity and legally defined definitions of what constituted toxicity. Thus, for example, the terms 'toxic' and 'very toxic' have been given quantitative definitions based on the LD_{50} in order that that substances can be labelled with these terms to provide some warning of danger for potential users. Of course, the older, less quantitative and more generalized usage has continued. This can lead to misunderstandings that may have serious consequences. For example, substances that are not labelled as 'toxic' may be assumed to be non-toxic and consequently used carelessly, with disastrous results. This is because the LD_{50}, the historic basis in law for classification as 'toxic', is, at best, an indication of the ability of a substance to cause death following short-term exposure. It tells us nothing about lethality of long-term exposure, probably a very common situation in the human context, or about other toxic effects that may be severely disabling but not immediately lethal.

Results of Misunderstanding of Concepts. The possible harmful effects of misunderstood toxicological terms on human health are fairly obvious to most people but the economic consequences tend to receive less attention by the general public. In Europe, the main driving force for development of

Concepts in Toxicology
By John H Duffus, Douglas M Templeton and Monica Nordberg

Published by the Royal Society of Chemistry, www.rsc.org

regulatory toxicology has been trade. It was already clear in the 1960s that trade in chemicals was being hindered by the variety of regulatory systems and hazard (toxicity) classifications applied in different countries around the world. This was an immediate problem for European nations seeking to establish free trade within the growing European Community, now the European Union (EU). The result was the 7th amendment to the Directive on Classification, Packaging and Labelling, that applied a common approach to toxicity assessment and subsequent labelling for all members of the European Community. Since most substances in commercial use had not been tested for toxicity, this was the start of a demand for toxicity testing that led subsequently to the Notification of New Substances Directive and which has culminated in the Registration, Evaluation, Authorisation and Restriction of Chemicals (REACH) Regulation that came into effect in 2008. This Regulation has made risk evaluation for toxicity and other hazards compulsory for all substances in use within the European Union. If acceptable, the substance will be authorized for marketing throughout the European Union. Without such authorization marketing is prohibited. The methods to be applied for toxicity testing and evaluation are described in detail in Technical Guidance Documents to ensure that the regulatory authorities in all the EU countries apply the same criteria and reach the same conclusions. This helps to ensure that trade is not impeded by contrary evaluations in the different countries of the EU. The downside is that any toxicological errors in the evaluation procedures are very difficult to change because of the inertia inevitable in such a system. Thus, erroneous evaluations can have a long-term effect on the availability of substances of value and may even result in the substitution of authorized toxic substances for less toxic alternatives that have not been authorized because the original evaluation was based on inadequate toxicology. REACH is available online at: <http://eur-lex.europa.eu/LexUriServ/LexUriServ.do?uri=OJ:L:2008:354:0060:0061%20:EN:PDF>.

The successor to REACH as state-of-the-art legislation is likely to be the Globally Harmonized System of Classification and Labelling of Chemicals (GHS), which is being developed by United Nations (UN) as a system to be adopted worldwide. The UN GHS is not a formal treaty, but instead is a non-legally binding international agreement. Therefore, countries (or trading blocks) must create local or national legislation to implement the GHS. The Organization for Economic Cooperation and Development (OECD) is developing proposals for classification criteria and labelling in the area of health and environmental hazards, at the request of the UN Sub-Committee of Experts on the GHS. A Task Force on Harmonization of Classification and Labelling has been established to coordinate the technical work carried out by the experts. The involvement of the OECD shows the importance that concepts in toxicology now have for world trade. An early and valuable consequence of this was the investment by the countries belonging to OECD of large amounts money to fund the development of Guidelines for Testing of Chemicals (including toxicity tests). These are now the 'gold standard' for toxicity tests and, where applied precisely as described, provide data that are accepted worldwide as the basis for decisions on risk management of chemicals in use. An important part of this has been the requirement for good laboratory practice and quality assurance. The current draft of GHS is available online at: http://www.unece.org/trans/danger/publi/ghs/ghs_rev02/02files_e.html. The current draft of the OECD Guidelines for Testing of Chemicals is available online at: <http://www.oecd.org/document_22/0,3343,en_2649_34377_1916054_1_1_1_1,00.html>.

In the negotiations for improved international understanding of chemical safety matters and implementation of harmonized laws, linguistic barriers lead to problems between nations and even between scientific disciplines in terms of comprehension and, therefore, in reaching agreement. This leads to considerable time being wasted at meetings in reaching a common perception of the nature and significance of problems and, hence, to difficulties in achieving a final agreement on acceptable solutions. The time that is wasted is also a waste of the money invested in the meeting and a waste of the valuable time of the participants. Hopefully, this book may help to reduce such waste.

Another consequence of misunderstood concepts in toxicology is misclassification of important substances, either prohibiting their use unnecessarily or permitting their use when the risk involved should have been perceived and avoided. A widely misunderstood concept has been ‘chemical speciation’. This concept is tacitly assumed for organic substances, which are all chemical species of carbon. Carbon compounds have never been given blanket toxicity classification on the basis of carbon being the main element present and, thus, fundamental to their structure. There is no subset of toxicology called ‘carbon toxicology’, whereas there is a subset called ‘metal toxicology’, even though different chemical species of metals have vastly different properties, just as carbon compounds do. The consequence of this misunderstanding of chemistry has been blanket classification of, for example, all nickel compounds as carcinogens, in the absence of evidence relating to most of the compounds, on the assumption that nickel cations will be released from all such compounds and that hydrated nickel cations are carcinogenic. There is little evidence to support this contention in spite of many related studies. As one might expect, there is no report of nickel alloys in any form, including stainless steel, coinage, dental and other surgical prosthetics, or even jewellery such as ear rings in intimate contact with the human body, having caused cancer, although allergic reactions have been reported for the last mentioned. By contrast, the most acutely toxic form of nickel is a gas, nickel tetracarbonyl, which is probably fatal through its carbonyl groups and not through the nickel atom, even if it does eventually release a free cation. Whether nickel tetracarbonyl is ever carcinogenic is a matter of dispute.

Structure of the Book

All concept entries start with related IUPAC-approved definitions from the IUPAC 'Gold Book'[1] or from the glossary of terms in toxicology published in *Pure and Applied Chemistry*.[2]

Particular attention is drawn to the concept diagrams presented in this introduction and at the beginning of each subset of concepts. No concept exists in isolation and each may cast some light on others. Although, for the purposes of quick reference, alphabetical arrangement of the terms is convenient and is used in all glossaries, related concepts may be separated widely by such organization. In this book we have adopted a different approach. We have arranged concepts in order of increasing complexity or of other logical relationship following construction of concept diagrams for this purpose. Each diagram attempts to show the concepts in logical order, beginning at the top of the diagram and proceeding to the bottom, following different routes as appropriate within any group of concepts. Each concept diagram is inevitably an oversimplification and the reader is advised to look at all the concept diagrams, where some concepts may occur more than once, to get a fuller picture of concept relationships. This should lead to a more profound understanding of the concepts described than is possible by considering each concept or concept diagram independently.

The concept diagrams themselves are arranged in the sequence shown in Figure 1. Firstly, from the top, there is a group of concepts related to fundamentals of toxicology. They have been divided into two subgroups. Subgroup A covers exposure and toxicity and Subgroup B covers hazard and risk assessment. The remaining concepts are arranged in groups according to the level of biological complexity involved. Thus, we start with the molecular and cellular level and proceed through organism and environmental levels, all of which apply directly to human toxicology. Two levels of toxicology shown in Figure 1 are not addressed specifically here, namely population toxicology and ecotoxicology. Both of these levels are related and introduce complexities that require a separate book to consider in depth.

Concepts in Toxicology
By John H Duffus, Douglas M Templeton and Monica Nordberg

Published by the Royal Society of Chemistry, www.rsc.org

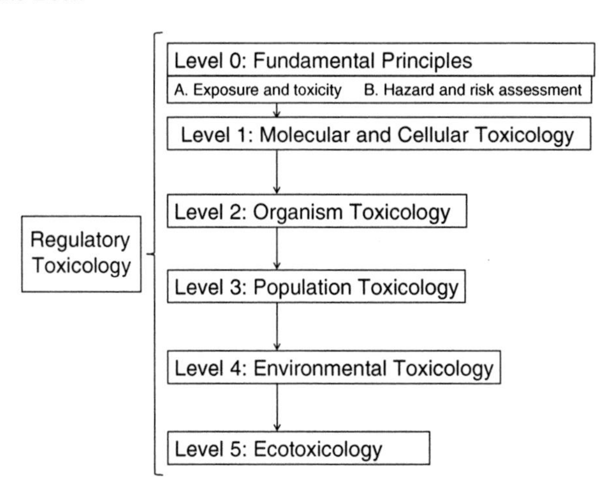

Figure 1 Conceptual subsets of toxicology arranged according to levels of systematic biological complexity, from the most fundamental principles at the top to the most complex at the bottom. Each subset can be further subdivided (see also Figures 2–5).

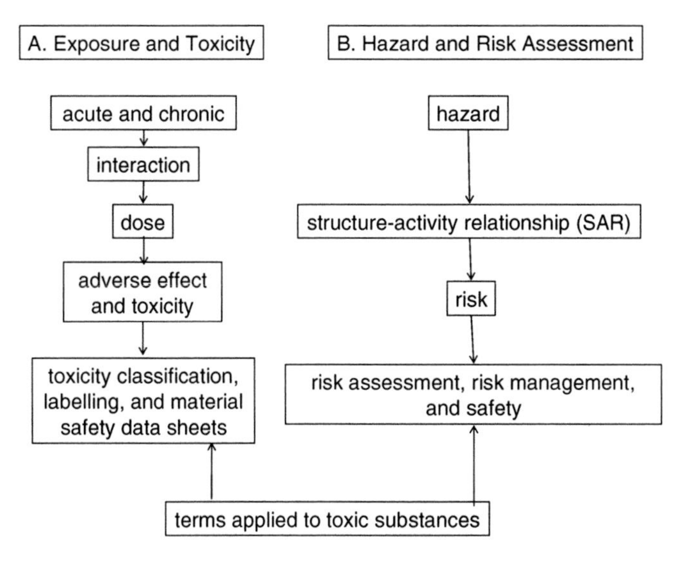

Figure 2 Concept group 1 (Level 0, Figure 1) – concepts of fundamental principle in toxicology – (Subgroup A) relating to exposure and toxicity; (Subgroup B) relating to hazard and risk assessment.

Fundamental Principles. Figure 2 shows the concepts of fundamental principle. There are two subgroups, A and B. The first, A, deals with exposure and toxicity. Before any toxic effect can occur there must be exposure to a toxic agent. Exposure may be acute or chronic and the resultant effects once a toxic dose is reached may also be classified as acute or chronic. Acute exposure may cause chronic effects and chronic exposure may cause an acute effect, manifesting itself in a short time once an internal dose threshold for effect is exceeded. The latter possibility is rare and normally chronic exposure is associated with chronic effect. Once tests of the effects of acute and chronic exposure have been completed and data collated, toxicity classification is possible, labelling may be defined and material safety data sheets (MSDS) may be compiled. This involves terminology that has specific toxicological meanings that must be understood.

The second subgroup of terms, B, relates to hazard and risk. The hazard (toxicity) of substances is related to their chemical structure and potential for biological activity. However, this potential will be realized only if exposure occurs and a toxic dose is reached in the organisms at risk. This may be highly improbable if exposure is kept to a minimum. If this condition is met, such improbability equates with low risk and a high degree of safety. Thus, there may be a highly toxic substance that is safe in use because exposure is restricted and risk is reduced to a minimum. Conversely, a substance of low toxicity may be used carelessly, leading to a high degree of exposure, high risk of toxicity, and a low safety level. The risk will be highly dependent on susceptibility of those at risk and thus on their genetics, considered in concept group 2. Minimization of risk by risk assessment and appropriate management must take this into account. It also requires effective communication and this brings us back to the terminology applied.

Molecular and Cellular Toxicology. Figure 3 shows the concepts related to molecular and cellular toxicology. Any substance can have effects on cell biology but this depends on chemical speciation. Chemical speciation depends upon the chemical structure of the substance of concern and the exact physicochemical properties of the environmental or physiological medium involved. Fundamentally, this determines the bioaccessibility of the substance. If the substance is not bioaccessible, it is not bioavailable and cannot harm directly any living organism. If the substance is both bioaccessible and bioavailable it may interact with and enter any cell with which it comes into contact. Interaction with a receptor on the outer surface of a cell may be sufficient to cause a toxic effect and may even be fatal in some circumstances. Interaction with a receptor on the outer surface may be the first stage of entering the cell (absorption). Entry may be by passive diffusion or by active transport (with input of metabolic energy), either of which may require a carrier. Once inside the cell, a substance may undergo biotransformation into derivatives that bind to a final receptor with toxic consequences. Biotransformation of the substance or its derivatives may depend upon the balance between aerobic and anaerobic metabolism and, hence, on the redox status of the cell. Reactive oxygen species (ROS) are often involved. Among the properties of ROS is mutagenicity and this may be associated with carcinogenicity. Mutagenicity alters the gene and protein complements of the cell, which are considered within the sciences of genomics and proteomics, respectively. Epigenetic changes may occur that affect the genome and the proteome.

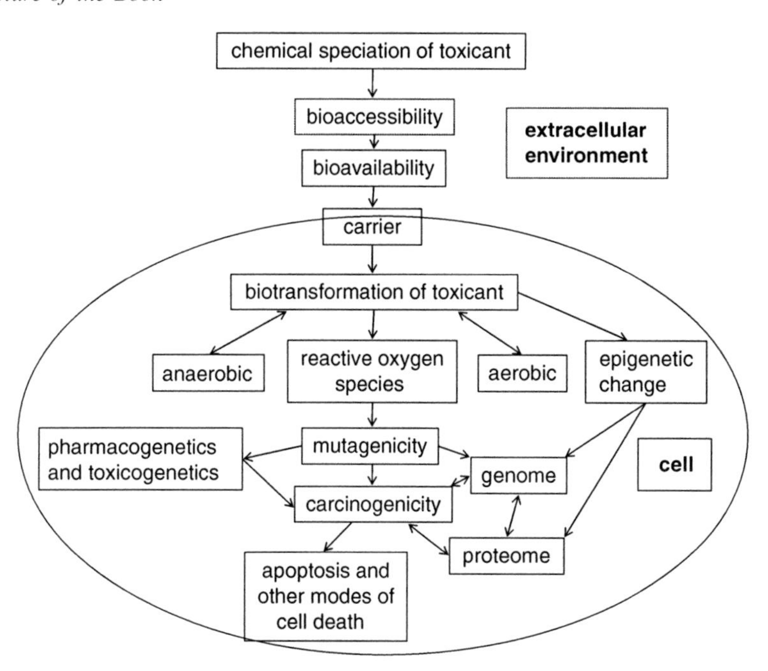

Figure 3 Concept group 2 (Level 1, Figure 1) – concepts in molecular and cellular toxicology.

Epigenetic changes may also lead to carcinogenicity and, in turn, to further mutagenic and epigenetic events. Differences in the genome (in pharmacogenetics and toxicogenetics) between individuals determine differences in the way in which they handle drugs and other toxicants. The ultimate outcome of these various processes may be apoptosis or other modes of cell death.

Whole Organism Toxicology. Figure 4 moves on to organism toxicology and shows the relationships between the terms relating to the effect of a substance on the whole organism. Cellular toxicology is of course fundamental to organism toxicology, but we have now moved on to consider cells that have differentiated and formed organs and tissues and the interactions between the differentiated cells as they function as part of the system that constitutes a multicellular organism. The key to protecting the whole organism is control of acute and chronic exposure, discussed above among the concepts of fundamental principle in toxicology. This requires a reference value that characterizes safe exposure. Most regulatory toxicologists now favour moving the reference value from the outdated LD_{50} or LD_{50} value to the benchmark concentration or dose (see below). The reference value is calculated differently depending upon whether it relates to a deterministic (non-stochastic) or stochastic effect. Whatever the exposure concentration may be, any substance of concern must usually be taken up by an exposed organism

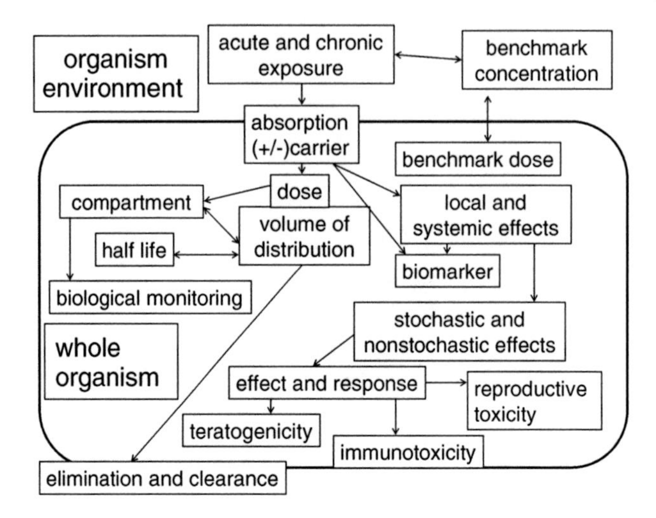

Figure 4 Concept group 3 (Level 2, Figure 1) – concepts in whole organism toxicology.

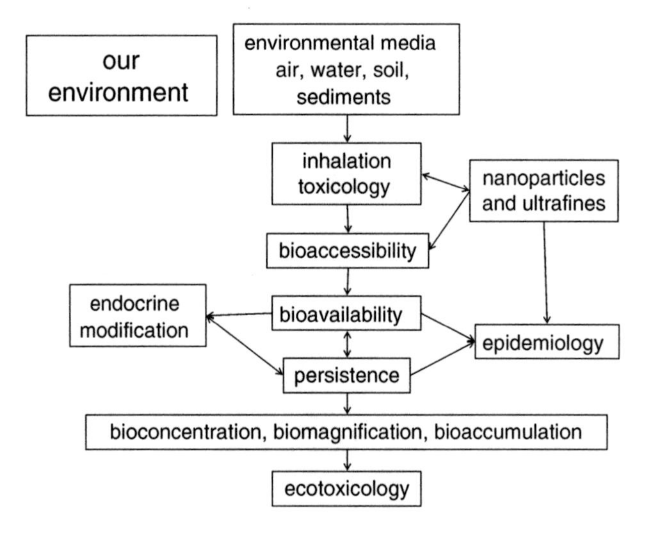

Figure 5 Concept group 4 (Level 4, Figure 1) – concepts in environmental toxicology.

to produce an absorbed dose before it can produce any local or systemic effect. Ultimately, this will result in a toxic effect and a response measured as the percentage of the population exposed to the given concentration showing the defined toxic effect. Once absorbed, the potentially toxic substance will be dispersed throughout the organism and a volume of distribution may be determined and described mathematically in terms of a compartment model. Subsequent elimination and clearance of the substance from the organism will help to define its biological half-life. Knowing the biological half-life is essential for the interpretation of the results of biological monitoring. Biological monitoring may be focused on selected biomarkers, measurable parameters that may be used indicators of exposure, dose or effect.

Environmental Toxicology. In Figure 5 we consider concepts in environmental toxicology. Environmental toxicology requires consideration of exposure through any or all of the environmental media: air, water, soil or sediment. Soil and sediment are composed of particles and there is increasing concern about inhalation toxicology of airborne particulates, especially nanoparticles and ultrafines. This brings us back to consideration of bioaccessibility and bioavailability, already mentioned above as fundamental concepts in toxicology. In the environmental context, persistence is a key concept in considering bioaccessibility and bioavailability. Persistence in the environment increases the likelihood of bioconcentration, biomagnification and bioaccumulation in food webs, making it a key factor in both risk assessment for effects on the human species and for assessment of ecotoxicology.

References

1. IUPAC, *Compendium of Chemical Terminology (the 'Gold Book')*, compiled by A. D. McNaught and A. Wilkinson, Blackwell Science, Oxford, 2nd edn, 1997. XML on-line corrected version, 2006–2008, created by M. Nic, J. Jirat and B. Kosata; updates compiled by A. Jenkins. Available at <http://goldbook.iupac.org>.
2. J. H. Duffus. M. Nordberg and D.M. Templeton, Glossary of terms used in toxicology, 2nd edn, *Pure Appl. Chem.*, 2007, **79(7)**, 1153. Available at <http://media.iupac.org/publications/pac/2007/pdf/7907x1153.pdf>

General Bibliography

Fundamental Toxicology, ed. J. H. Duffus and H. G. J. Worth, Royal Society of Chemistry, Cambridge, 2nd edn, 2006.

Principles and Methods of Toxicology, ed. A. W. Hayes, CRC Press, Boca Raton, FL, 5th edn, 2007.

Casarett & Doull's Toxicology – The Basic Science of Poisons, ed. C. D. Klaasen, McGraw-Hill, New York, 7th edn, 2008.

Handbook on the Toxicology of Metals, ed. G. F. Nordberg, B. A. Fowler, M. Nordberg and L. T. Friberg, Elsevier, Amsterdam, 3rd edn, 2007.

M. Nordberg, D. M. Templeton, O. Andersen and J. H. Duffus, Glossary of terms used in ecotoxicology (IUPAC Recommendations 2009), *Pure Appl. Chem.*, 2009, **81**, 829.

OECD, *Guidelines for Testing of Chemicals*, OECD, Paris, 1982 onward.

Concept Groups

Concept Group 1. Concepts Applying to Fundamental Principles of Toxicology

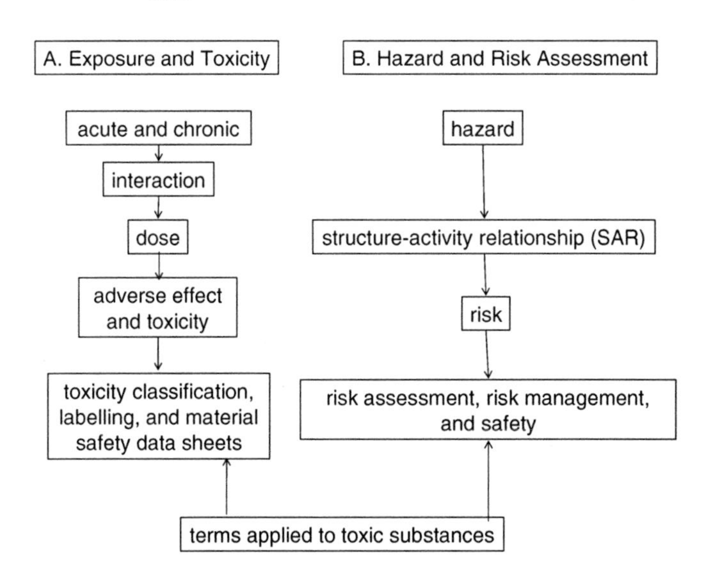

Subgroup A – Exposure and Toxicity

Concepts in Toxicology
By John H Duffus, Douglas M Templeton and Monica Nordberg

Published by the Royal Society of Chemistry, www.rsc.org

1.1 Acute and Chronic

A. Acute

acute
1 Of short duration, in relation to exposure or effect. In experimental toxicology, acute refers to studies where dosing is either single or limited to one day, although the total study duration may extend to two weeks.
2 In clinical medicine, sudden and severe, having a rapid onset.
Antonym: chronic.
acute effect
Effect of finite duration occurring rapidly (usually in the first 24 h or up to 14 d) following a single dose or short exposure to a substance or radiation.
acute exposure
Exposure of short duration.
Antonym: chronic exposure.
acute toxicity
1 Adverse effects of finite duration occurring within a short time (up to 14 d) after administration of a single dose (or exposure to a given concentration) of a test substance or after multiple doses (exposures), usually within 24 h of a starting point (which may be exposure to the toxicant, or loss of reserve capacity, or developmental change, *etc.*).
2 Ability of a substance to cause adverse effects within a short time of dosing or exposure.
Antonym: chronic toxicity.

In toxicology, 'acute' is a word that is used in combination with exposure, toxicity and effect. Acute exposure is a single or very short-lasting dosing by any route. Talking about acute toxicity addresses adverse effects (further discussed below), *i.e.* harmful effects, unwanted negative effects that occur immediately after or within a short time after administration of a single dose of a substance, or following short exposure or concurrently with continuous exposure, or recurrently following shortly after multiple doses. 'Short' implies a time of 24 h or less. Some effects considered to be acute can occur up to as long as 96 h after exposure. Uraemia can be an acute effect, but it takes almost 96 h to see such an outcome. In toxicity testing, it is most important to be aware of this in order not to draw any false conclusions from animal studies with agents that cause such an acute effect. Acute effects usually occur or develop rapidly after a single exposure. However, acute effects can also appear immediately after, or during, repeated or prolonged exposure.

Acute Toxicity. Historically, an important aspect of acute toxicity has been the identification of the lethal dose or exposure that kills an organism after a short exposure or a single dose. This has been established by a test in which selected organisms are exposed to a series of increasing dose levels until a dose is reached at which all the organisms die. For regulatory purposes, to permit extrapolation to humans, it is usually performed with at least two mammalian species. From such tests, the LD_{50} for the test species has been derived and used for the classification of the toxicity of chemicals to humans. Such tests involved killing large numbers of animals to obtain a toxicity classification based on lethality. However, this classification tells us nothing about sublethal effects such as immunotoxicity or teratogenicity. This situation was clearly unsatisfactory and so acute toxicity testing is now designed in such a way as to obtain maximum information about all aspects of acute toxicity using the minimum number of animals.

In Europe, classification of new chemicals for toxicity is no longer based on the LD_{50}. The tests used for this purpose are based on survival rather than on lethality. For example, the method of fixed-dose testing is usually limited to a maximum dose of 500 mg per kg body weight. If five males and five females exposed to a dose of this magnitude survive with no evidence of toxicity, the chemical tested need not be classified as toxic. Toxicity classifications based on this approach can provide a similar classification of toxicity to the old LD_{50} system but with a huge reduction both in the number of animals used and in animal suffering compared to the traditional LD_{50} tests. Another approach to reducing the numbers of animals used is the up-and-down procedure, which also produces a value approximating to the LD_{50}. This procedure uses sequential dosing together with sophisticated computational methods. It provides a point estimate of the LD_{50} while achieving significant reductions in animal use.

Since chronic toxicity testing (see below) is expensive and labour-intensive, there is a great need to replace it where possible with shorter-term predictive acute tests and early identification of biomarkers of toxicity. This has been possible to some extent with carcinogenicity. In the past, a cancer study was designed to expose animals to the toxicant and to follow the animals during their lifetime. Each animal upon death was examined for occurrence and localization of tumours in the body. Since it is expensive to maintain animals over long periods, the need for new tools to identify carcinogens is clear. Many cancers start with mutations or chromosome damage, and this can be assessed with short-term tests such as the Ames test or the host-mediated (Legator) test. The Ames test is based on reversal of a point mutation in a *Salmonella* strain, which makes it unable to synthesize the amino acid histidine. Back-mutation can be detected by growth of the bacteria in a histamine-depleted medium. Rat liver microsomes are included in the test medium to simulate the metabolic activation of organic compounds that may take place in the intact animal. The host-mediated test looks for chromosome changes *in vitro* and (or) *in vivo*, including chromosome breaks and sister chromatid exchanges, in microbial cells introduced (*e.g.* by intravenous injection) into a host animal. The host animal receives the test compound orally and therefore acts as a source of chemical metabolism, distribution and excretion. Another whole animal test involves looking for the production of micronuclei in animals exposed to possible carcinogens. The micronucleus test is less sensitive than bacterial tests but is a more realistic measure of likely chromosomal damage in mammals at risk.

It is also possible to test quickly for mutagenicity and the possibility of associated carcinogenicity by adding suspect substances to cell cultures and looking for chromosome damage and cell transformation. Another approach to carcinogenicity testing is to apply the substances to tissues in culture and/or to genetically compatible transplants and similarly assess the changes that occur.

While the acute tests for mutagenicity give a quick indication of the mutagenic potential of substances tested, it must be emphasized that the effects observed may not necessarily extrapolate to the intact organism. The bacterial strains used in the Ames test have been selected for the absence of DNA (deoxyribonucleic acid) repair mechanisms so that they are much more sensitive to mutagenicity than any normal organism. Cultured cells and tissues lose differentiated properties and are abnormal in this way. Dedifferentiated cells tend to divide more rapidly than normal, and this may facilitate chromosomal damage. In assessing carcinogenicity, it must be remembered that not all mutations lead to cancer nor are all cancers the result of mutations. Thus, while these acute tests may indicate the possibility of carcinogenicity, they are not sufficient to prove it and can be regarded only as screening tests to select substances for further study in this regard.

B. Chronic

chronic
Long-term (in relation to exposure or effect).
1 In experimental toxicology, chronic refers to mammalian studies lasting considerably more than 90 days or to studies occupying a large part of the lifetime of an organism.
2 In clinical medicine, long-established or long-lasting.
Antonym: acute.
chronic effect
Consequence that develops slowly and/or has a long-lasting course: may be applied to an effect that develops rapidly and is long-lasting.
Antonym: acute effect.
Synonym: long-term effect.
chronic exposure
Continued exposures occurring over an extended period of time, or a significant fraction of the test species' or of the group of individuals', or of the population's lifetime.
Antonym: acute exposure.
Synonym: long-term exposure.
chronic toxicity
Adverse effects following chronic exposure. Effects which persist over a long period of time whether or not they occur immediately upon exposure or are delayed.
Antonym: acute toxicity.

Chronic effects usually occur after repeated or prolonged exposures. However, chronic effects can also occur after single exposure if they develop slowly or are long-lasting. They are often irreversible. Chronic effects may follow accumulation of a toxic substance or of metabolites formed by biotransformation of the administered substance. They may also be the result of cumulative irreversible effects of toxicants.

Chronic Toxicity. Chronic toxicity usually results in a progressive loss of organ function, *e.g.* increasing liver damage following regular ingestion of ethanol. For humans, a particularly serious example of chronic toxicity may be the gradual loss of brain cells due, for example, to excessive exposure to ethanol or other neurotoxic agents. Brain cells do not divide and cannot be replaced once they are lost. Because we have a large reserve capacity of such cells their gradual loss may not be apparent, but this added to the normal loss associated with aging may result in premature dementia and related adverse effects.

For the toxicologist, a particular problem arises when the dose or exposure is low or the effect develops a very long time after exposure as may happen with cancer. In these circumstances, it is very difficult to attribute a cause to the delayed effect. It is also difficult to test substances for such effects. Cancer in humans may take up to 40 years to develop after exposure to a carcinogen. Our normal test animals, rats and mice, have life spans of about 2 years and 18 months, respectively. To cause malignant tumours within such a short time, very large doses of suspect carcinogens must be applied. Thus, test doses are much higher than those to which humans may ever be exposed and may therefore overwhelm metabolic defence mechanisms that work well within normal human exposure ranges.

Subchronic (sometimes referred to confusingly as subacute) toxicity refers to the adverse effects observed when animals are administered a toxicant over time, as a result of repeated daily dosing of a chemical, or exposure to the chemical, for a significant part of an organism's lifespan (usually not exceeding 10%). Observations of acute and subchronic toxicity indicate what the critical (target) organ and the critical effect are. With experimental animals, the subchronic period of exposure may range from a few days to 6 months. The terms 'subchronic' and 'subacute' suffer from many variations in their usage and are best avoided. It is better to replace them by giving precise definition of the times of administration and observation. Subchronic testing has usually been limited to 90 days. Chronic toxicity testing should be over the lifetime of the organism, which means 1.5–2 years in the case of the mouse or the rat.

Chronic toxicity testing in rodent and non-rodent species identifies not only general toxicity but also aspects of mutagenicity, carcinogenicity and reproductive toxicity (in rats and rabbits), including specific effects on the reproductive organs, teratogenicity and reproductive toxicity. Some strains of mice, for example, have different frequencies of naturally occurring effects. Most chronic studies are carried out with at least two animal species, usually rats and a non-rodent species such as dogs or primates. For cancer testing, it is important to choose a species known to have an intrinsically low frequency of tumours. For example, the Syrian golden hamster has a low background frequency of tumours in the trachea and lung and thus may be chosen to test for carcinogens suspected to target these organs.

Currently there is great activity in developing alternatives to chronic animal testing, *e.g.* the use of stem cells, tissue culture and *in silico* methods.

Further Reading

E. Walum, Acute oral toxicity, *Environ. Health. Perspect.*, 1998, **106**(Suppl 2), 497. Available at <http://www.ehponline.org/realfiles/members/1998/Suppl-2/497-503walum/walum.html>.

N. H. Wilson, J. F. Hardisty and J. R. Hayes, Short-term, subchronic, and chronic toxicology studies, in *Principles and Methods of Toxicology*, ed. A. W. Hayes, CRC Press, Boca Raton, FL, 5th edn, 2007.

1.2 Interaction

additive effect
Consequence that follows exposure to two or more physicochemical agents that act jointly but do not interact: the total effect is the simple sum of the effects of separate exposures to the agents under the same conditions.
potentiation
Dependent action in which a substance or physical agent at a concentration or dose that does not itself have an adverse effect enhances the harm done by another substance or physical agent.
synergism (in toxicology)
Pharmacological or toxicological interaction in which the combined biological effect of two or more substances is greater than expected on the basis of the simple summation of the toxicity of each of the individual substances.
antagonism
Combined effect of two or more factors which is smaller than the solitary effect of any one of those factors. In bioassays, the term may be used when a specified effect is produced by exposure to either of two factors but not by exposure to both together.

When an organism is exposed to two or more substances that produce a particular physiological effect these substances may or may not interact. If there is no interaction the effects would be strictly additive; this is intuitively obvious, and the *IUPAC Glossary of Terms Used in Toxicology*, 2nd Edition, cited in the Introduction, defines additive effect accordingly. The substances would in general each show a dose response–effect individually, and the effects would be strictly additive at any combination of concentrations. This shifts our focus from the substances to the effect: additivity or other descriptors of interaction do not describe the substances themselves, but rather the effects that they elicit. A corollary is that to assert that two substances behave in an additive fashion requires that no statistically significant difference can be demonstrated between measurements made upon exposure to the substances together compared to the sum of the individual exposures.

Additivity. From the above considerations, additivity is strictly the result of substances acting together but independently in the production of a toxic effect. The following are therefore characteristics of additivity:

1. Additivity (or lack thereof) refers only to what can be measured. Therefore, it refers only to specific effects. Two substances that may be additive with respect to a certain effect may be non-additive with respect to other effects. For example, two drugs might be strictly additive with respect to an effect on blood pressure, but have non-additive effects on liver function. So-called drug–drug interactions often refer to non-additivity with respect to side effects.
2. Additivity may occur at some concentration ratios and not others. Because any substance may be non-toxic at some levels and toxic at others (remember Paracelsus, it is only a question of the dose), non-additive effects may be observed only when one component reaches a critical, threshold concentration. For instance, two anticancer drugs might have additive effectiveness until one reached a threshold concentration for suppressing angiogenesis, at which point the effectiveness of the other might increase based on an ability to target hypoxic tissue.
3. Additivity, in the strictest sense, is in general probably not the norm. The complexity of biological systems is such that multiple effects will probably occur with any bioactive agent, and any overlap with the effects of a second agent will produce candidate effects for non-additivity.

Non-additive Interactions. When two or more substances are related through a common toxic, therapeutic or other biological effect, and yet their effect is non-additive with respect to a measured parameter, this interaction may be described by one of several different terms. When the effect of one substance is diminished by the presence of a second, the situation is fairly straightforward. Antagonism is defined above as the 'combined effect of two or more factors which is smaller than the solitary effect of any one of those factors', with the added comment that, in bioassays, the term antagonism 'may be used when a specified effect is produced by exposure to either of two factors but not by exposure to both together'. In this case, it is not necessary to specify which of the two factors is decreasing the activity of the other; both are assumed to have a certain activity, which when they are present together is less than additive. This distinguishes antagonism from inhibition, where one substance may or may not elicit an effect common with another, but is nevertheless capable of antagonizing that effect. When exposure is to more than two substances, *i.e.* to mixtures, outcomes are often difficult to predict.

Perhaps a more challenging distinction is between potentiation and synergism, situations where the measured effect of two or more agents is greater than that attributed to either alone. The parent IUPAC Glossary defines potentiation as, 'Dependent action in which a substance or physical agent at a concentration or dose that does not itself have an adverse effect enhances the harm done by another substance or physical agent', and synergism as (in toxicology) 'Pharmacological or toxicological interaction in which the combined biological effect of two or more substances is greater than expected on the basis of the simple summation of the toxicity of each of the individual substances'. In essence, potentiation refers to an effect of substance A to increase the effect of B, when A itself does not cause the same effect as B, whereas synergism means that A and B share a common effect, which is greater than additive when both are present.

Maintaining this distinction between potentiation and synergy may not be very useful. One argument would be that if B has no influence on an effect that is elicited by A, but when present increases the effect of A, then it would be said to potentiate the effect of A. On the other hand, if B has an effect in common with A, and both when present together give an effect that is greater than the sum produced by both alone, then we would call that synergism. But this distinction is not straightforward, because it is often experimentally difficult to determine whether the very effect measured may be elicited by one substance only in the presence of the other. Suppose that A and B do not interact but elicit a common effect, E, with contributions from isolated exposures of E_A and E_B, respectively, such that when given together $E_{AB} = E_A + E_B$ (additivity). Now suppose, on the other hand, that B can influence the effect of A (E_A), so that this effect of A has a 'pure' component, E_{Aa}, and a component dependent on the presence of B, E_{Ab}. Then $E_A = E_{Aa} + E_{Ab}$. Ignoring for the moment any effect of A on E_B, we can write $E_{AB} = E_{Aa} + E_{Ab} + E_B$. If $E_{Ab} = 0$, we have additivity. If $E_{Ab} < 0$, we have antagonism. But if $E_{AB} > E_A + E_B$, is this because $E_{Ab} > 0$ with $E_B = 0$ (potentiation) or with $E_B > 0$ (synergy)? The decision may be difficult experimentally, when one considers that A may also have an influence on E_B (*i.e.* $E_{AB} = E_{Aa} + E_{Bb} + E_{Ab} + E_{Ba}$). Even if B has no effect in isolation, E_B may be non-zero only in the presence of A.

Finally, notably, some textbooks of pharmacology recommend against the use of potentiation, referring to any increase above additivity ($E_{AB} > E_A + E_B$) as synergism, regardless of whether either E_A or E_B is zero.

Further Reading

EPA Mixtures Guidance, 2001. Available at <http://www.epa.gov/ncea/raf/pdfs/chem_mix/chem_mix_08_2001.pdf>.

U.S. Department of Health and Human Services Public Health Service, Agency for Toxic Substances and Disease Registry, Division of Toxicology, *Guidance Manual for the Assessment of Joint Toxic Action of Chemical Mixtures*, May 2004. Available at <http://www.atsdr.cdc.gov/interactionprofiles/ipga.html>.

C. Winder and A. Zarei, Incompatibilities of chemicals, *J. Hazard. Mater.*, 2000, **A79**, 19.

1.3 Dose

dose (of a substance)
Total amount of a substance administered to, taken up, or absorbed by an organism, organ, or tissue.

In common pharmaceutical usage, the term 'dose' is applied to the amount of medication taken by a patient at any one time. In the IUPAC definition, this is covered by the term 'administered' but, ideally, the toxicologist wants to know the amount of a substance 'taken up, or absorbed', in the same definition. In other words, the toxicologist would like to know the amount available for interaction with metabolic processes or biologically significant receptors after crossing the relevant biological boundary (epidermis, gut, respiratory tract, cell membrane). This 'absorbed dose' is the amount crossing a specific absorption barrier and would be best defined in practice if it could be referred specifically to the target organ but this is rarely possible.

Determination of Dose. Scientifically, it would be best if the dose were always expressed in molar terms (amount of substance) so that comparison could be made between the numbers of molecules involved and even related to numbers of receptor molecules. In practice, units of mass are more common in relation to prescription drugs.

A major problem for toxicologists is the relationship between exposure and internal dose. For a given exposure in a given medium, uptake of a substance into the body, and hence internal dose, varies from individual to individual depending upon physiology, behaviour and the presence of other substances that may prevent or enhance uptake. Because of our physiology, when air is contaminated with pollutants, people are advised to minimize exercise as this is associated with rapid deep breathing, resulting in greater uptake of the pollutants from the air than would occur in someone at rest exposed to the same air concentration. Because of their behaviour, soil contaminants at a given concentration may be absorbed more by children who put soil in their mouths than by adults who do not behave in this way. Finally, interactions may occur preventing or facilitating uptake. For example, the presence of sufficient calcium ions in water prevents uptake of lead ions, minimizing the internal dose associated with a given lead ion concentration.

To determine the internal dose of a given chemical, analysis of tissues and body fluids can be carried out. In determining the internal dose, analysis is aimed at measuring amounts of the substance itself, and (or) of its metabolites. Since the internal dose is defined as the total amount of a substance absorbed, measurement of the substance of concern should be repeated over the full period of exposure and the measurement results should be integrated over this time.

Further Reading

Principles and Methods of Toxicology, ed. A. W. Hayes, CRC Press, Boca Raton, FL, 5th edn, 2007.

1.4 Adverse Effect and Toxicity

adverse effect
Change in biochemistry, morphology, physiology, growth, development, or lifespan of an organism which results in impairment of functional capacity or impairment of capacity to compensate for additional stress or increase in susceptibility to other environmental influences.
toxicity
1 Capacity to cause injury to a living organism defined with reference to the quantity of substance administered or absorbed, the way in which the substance is administered and distributed in time (single or repeated doses), the type and severity of injury, the time needed to produce the injury, the nature of the organism(s) affected, and other relevant conditions.
2 Adverse effects of a substance on a living organism defined as in **1**.
3 Measure of incompatibility of a substance with life: this quantity may be expressed as the reciprocal of the absolute value of median lethal dose ($1/LD_{50}$) or concentration ($1/LC_{50}$).

Living organisms have evolved to adapt to a range of environmental conditions and to change in response to environmental changes, including chemical changes. Such changes in organisms in response to the environment constitute effects. These effects may be beneficial (as perhaps with essential nutrients), neutral or harmful. Harmful effects of a substance, related to dose, define its toxicity. Extreme damage is easy to identify, but the toxicologist wants to define the earliest signs of harm and this is not so easy. If the effect resulting from exposure to a potentially toxic substance is small it may be within the normal range of physiological variation required for life to adapt and cause no harm. In that case, it is not an adverse effect. In contrast, it may have a small effect that causes no immediate harm but may contribute to harm in future if the organism lives long enough. For example, lead may replace calcium in bone with no immediate effect but may accumulate there with time to cause harm during illness, pregnancy or old age. Thus, the apparently clear definition of an adverse effect may become difficult to apply in practice.

Once an adverse effect has been identified, it is important to know whether it is reversible or irreversible. It may be possible for an organism to recover completely from a reversible effect, but irreversible effects can accumulate with time and repeated exposures.

In assessing the consequences of adverse effects, the organ most affected, the critical organ, is a key factor, together with the dose–effect relationship. If one knows the initiating reaction for the adverse effect, this may help not only to assess the likely outcome but also to suggest treatment to alleviate the effect, *e.g.* by blocking the active site for the toxicant on a receptor molecule.

Distinguishing Between Adverse and Non-adverse Effects. The simplest definition of an adverse, or 'abnormal', effect experimentally is a measured effect that is outside the 'normal' range. The normal range is usually defined on the basis of values observed in a group of presumably healthy individuals, and expressed statistically as a range representing the 95% confidence limits (CLs) of the mean or, if the mean has been determined on the basis of a very large sample, the 95% limits will be equal to $\mu \pm 1.96\sigma$, where μ (mu) and σ (sigma) are the population values of the mean and the standard deviation (SD), respectively. An individual with a measured value outside this range may be either genuinely 'abnormal' or one of a small group of 'normal' individuals who have extreme values. This distinction between 'normal' and 'abnormal' values based on statistical considerations may be used as a criterion for adverse effects, if the exposed population consists of adult, generally healthy individuals, subject to periodical medical examination, such as workers. In these circumstances, departures from normal values associated with a given exposure can be considered as adverse effects, if the observed changes are:

1. statistically significant ($P < 0.05$) in comparison with a control group, and outside the limits ($\mu \pm 2\sigma$) of generally accepted 'normal' values;
2. statistically significant ($P < 0.05$) in comparison with a control group, but within the range of generally accepted normal values, provided such changes persist for a considerable time after the cessation of exposure;
3. statistically significant ($P < 0.05$) in comparison with a control group, but within the normal range, provided statistically significant departures from the generally accepted normal values become manifest under functional or biochemical stress.

The Student's *t*-distribution is usually important and can sometimes be applicable for datasets with a small number of components.

The preceding statistical considerations will not be appropriate if the available data are nonparametric, *i.e.* not consistent with a Gaussian (normal) distribution. For nonparametric data, the median replaces the mean as a measure of the central value. Two of the simplest nonparametric procedures are the sign test and median test. The sign test can be used with paired data to test the hypothesis that differences are equally likely to be positive or negative (or, equivalently, that the median difference is 0). The median test is used to test whether two samples are drawn from populations with the same median. The median of the combined dataset is calculated, and each original observation is classified according to its original sample (A or B) and whether it is less than or greater than the overall median. The chi-square test for homogeneity of proportions in the resulting 2-by-2 table tests whether the population medians are equal. The major disadvantage of nonparametric techniques is clear from the name. Because the procedures are nonparametric there are no parameters to evaluate and it becomes more difficult to make quantitative statements about the actual difference between populations. For example, when the sign test says two treatments are different, there is no confidence interval and the test does not say by how much the exposures differ. However, it is sometimes possible with the right software to compute estimates (and even confidence intervals) for medians and differences between medians. The second disadvantage of nonparametric procedures is that they throw away information. The sign test uses only the signs of the observations. Ranks preserve information about the order of the data but discard the actual values. Because information is discarded, nonparametric procedures can never be as powerful (able to detect existing differences) as their parametric counterparts when parametric tests can be used.

The statistical definition of adverse effects is likely to be inappropriate for a population that includes groups that may be specially sensitive to environmental factors, particularly the very young, the very old, those affected with disease and those exposed to other toxic materials or stresses. For such a population, which is the norm for humans, it is practically impossible to define 'normal' values, and any observable biological change may be considered as an adverse effect under some circumstances. Thus, it is important to establish criteria for adverse effects based on biological considerations as well as on statistically significant differences relating to an unexposed population (control group).

There are no generally agreed-upon biological criteria, and so ultimately the decision on what is an adverse effect tends to depend on experience and expert judgment. It may nevertheless be useful to give examples of such criteria, illustrating at the same time the difficulties in applying these criteria. One approach is to try to define which effects are non-adverse and to eliminate them from further consideration. Non-adverse effects have been defined negatively as the absence of changes in morphology, growth, development and life span. In addition, non-adverse effects are those that do not result in impairment of the capacity to compensate for additional stress. Non-adverse effects should be reversible following the end of exposure without any detectable reduction in the ability of the organism to maintain homeostasis, and should not enhance its susceptibility to the harmful effects of other environmental influences. Thus, in contrast, adverse effects may be defined as changes that:

1. follow single, intermittent or continued exposure and that result in loss of functional capacity (as determined by anatomical, physiological, and

biochemical or behavioural parameters) or in a decrease in the ability of the organism to compensate for additional stress;
2. are irreversible during exposure or following the end of exposure if such changes cause detectable loss in the ability of the organism to maintain homeostasis;
3. enhance the susceptibility of the organism to damaging effects of other environmental influences.

Application of the above criteria may be based on overt pathology (*e.g.* inflammation, necrosis, hyperplasia) or on metabolic and biochemical changes. Such changes may be considered to be adverse if, for example, enzymes that have a key significance in metabolism are inhibited or if there are changes in subcellular membranes (*e.g.* lysosomal membranes) resulting from the action of toxic substances. However, such changes may be within the limits of homeostasis or have no resultant pathology. Thus, the degree of change becomes a crucial measurement. One may debate, for example, what percentage inhibition of an enzyme must occur before harm results to an organism. Differentiating between 'non-adverse' and 'adverse' effects requires considerable knowledge of the reversible changes, which may be part of normal homeostasis. It also requires understanding of the subtle changes from 'normal' physiology and morphology, which may alter biological properties such as the ability to adapt to stress, and life expectancy. In considering possible harm to humans, the psychological and behavioural changes accompanying small effects on the nervous system may be particularly important. Such changes may follow exposure to certain metals and their derivatives. Examples of substances that may cause these changes are lead(II) ions and methylmercury. Special attention must be paid to their effects in children.

Of course, none of the above considerations or criteria can be applied if the data available for analysis are inadequate. One must be sure that the animal data have been obtained under test conditions that would show up all the effects that might occur. Key considerations here are the number of animals studied and the time and environmental conditions of exposure and observation. Too few organisms or too short a study may result in lack of statistical power to identify effects that occur in only a small susceptible group in the population. Some effects may occur under conditions of environmental stress that are rarely simulated in toxicity testing. It may be that more flexibility of test conditions should be introduced to simulate specific conditions associated with high risk of adverse effects on humans to provide the information we need to avoid toxic exposures to certain chemicals.

Further Reading

IPCS, *Principles and Methods for Evaluating the Toxicity of Chemicals, Part 1, Environmental Health Criteria 6*, WHO, Geneva, 1978. Available at <http://www.inchem.org/>.

1.5 Toxicity Classification, Labelling and Material Safety Data Sheets

toxicity
1 Capacity to cause injury to a living organism defined with reference to the quantity of substance administered or absorbed, the way in which the substance is administered and distributed in time (single or repeated doses), the type and severity of injury, the time needed to produce the injury, the nature of the organism(s) affected, and other relevant conditions.
2 Adverse effects of a substance on a living organism defined as in **1**.
3 Measure of incompatibility of a substance with life: this quantity may be expressed as the reciprocal of the absolute value of median lethal dose ($1/LD_{50}$) or concentration ($1/LC_{50}$).
safety data sheet
Single page giving toxicological and other safety advice, usually associated with a particular preparation, substance, process, use pattern, or exposure scenario.

Most countries have legislation requiring that all chemicals should have a clear label to indicate their identity, hazardous properties, and safety precautions. The label should draw attention to the inherent danger to persons handling or using the chemical. Symbols and pictograms (Figure 6) have been established

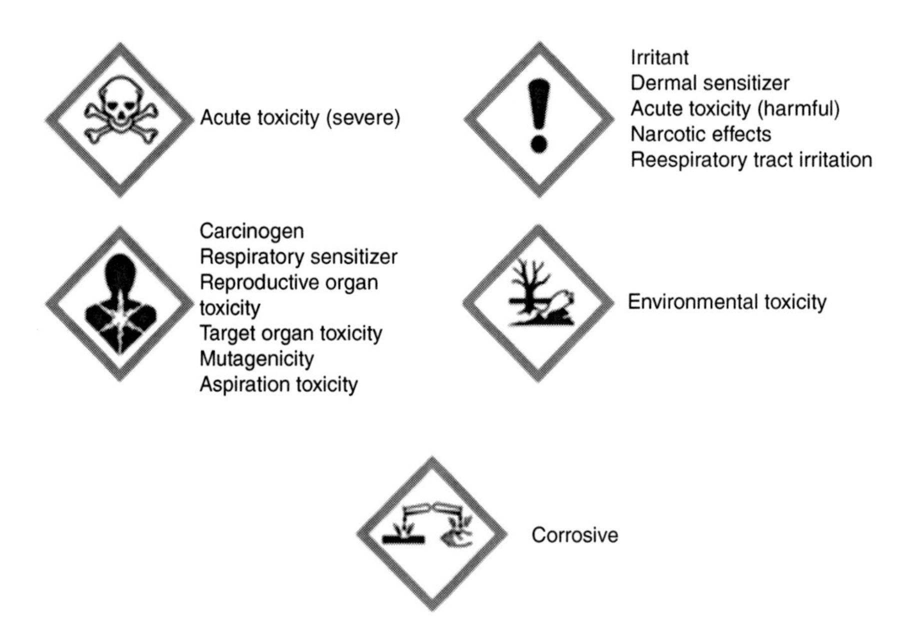

Figure 6 Global Harmonization System (GHS) pictograms used on labels to warn of properties hazardous to health.

for various hazard categories. These symbols or pictograms are an integral part of the label and give an immediate idea of the types of hazard that the substance or the preparation may cause.

The move to establish a Globally Harmonized System (GHS) for toxicity classification and labelling became fully established at the World Summit on Sustainable Development (Johannesburg, South Africa) in August–September 2002, when governments adopted the Johannesburg Plan of Implementation, which in para. 23c:

> encourages countries to implement the new globally harmonized system for the classification and labelling of chemicals as soon as possible with a view to having the system fully operational by 2008

Several countries have already implemented this system fully.

Notably, classification and labelling for supply of chemicals may be different from that for transport of chemicals. Classification and labelling for transport is dealt with by United Nations Organizations while that for supply is regulated at European Union level for the component nations and at national level otherwise.

Risk and Safety Phrases. There is an increasing international use of standard risk and safety phrases on labels. In the course of time, some phrases have been removed as improved phrases are introduced. Thus, there are some gaps in the numbering.

It should be emphasized that the so-called 'risk phrases' are really 'hazard phrases' since they give no indication of the probability of harm. This has been recognized in the development of the Globally Harmonized System (GHS) and the corresponding label components are called Hazard (H) statements. Accurate determination of the probability of harm associated with defined circumstances is the key to risk assessment (Section 1.2.4), and is dependent on the probability of exposure and resultant harm. If there is no exposure to a hazard, there is no risk. Examples of the so-called risk phrases (**R**) relating to toxicity are:

- **R 20** Harmful by inhalation
- **R 21** Harmful in contact with skin
- **R 22** Harmful if swallowed
- **R 23** Toxic by inhalation
- **R 24** Toxic in contact with skin
- **R 25** Toxic if swallowed
- **R 26** Very toxic by inhalation
- **R 27** Very toxic in contact with skin
- **R 28** Very toxic if swallowed

These may be used in combinations, *e.g.* R 20/21/22. 'Harmful' is the lowest level of toxicity classification. It implies that a large or continuing exposure is required to produce any harmful effect. 'Toxic' or 'very toxic' implies that a small exposure can cause harm fairly quickly. Thus, substances labelled 'Toxic' or 'Very Toxic' must be handled with great care. Substances labelled 'Harmful' must also be handled with care but there is less risk of harmful effects following a small exposure.

The standard safety phrases (S) give advice on the precautions necessary in handling chemicals. Indeed, in the GHS, they are called Precautionary (P) statements. In effect, these phrases describe how to reduce the risk in handling the substances to which they apply. Such precautions do not, of course, reduce the intrinsic hazards of the relevant substances. For example:

- **S 1** Keep locked up
- **S 2** Keep out of the reach of children
- **S 3** Keep in a cool place
- **S7** Keep container tightly closed
- **S9** Keep in a well-ventilated place
- **S14** Keep away from (incompatible materials to be indicated by the manufacturer)
- **S49** Keep only in the original container

As with the risk phrases, these are often used in combinations.

Labelling Requirements. In general, a label must clearly show:

1. the trade name;
2. the name and the address, including telephone number, of the manufacturer, the importer or the distributor;
3. the chemical name of the substance (in the case of a preparation, the chemical names of the hazardous components);
4. danger symbols;
5. risk phrases (R-phrases);
6. safety phrases (S-phrases);
7. the quantity of the contents of the package or container.

Labels should be in the national, official language(s) of the country where the substance is to be used.

A label should show the chemical names of substances present that are the most serious hazards. In some cases, the list of names may be quite long; for example, all cancer-causing substances in a preparation must be identified and the corresponding R- and S-phrases must be given on the label.

If the preparation contains one or more of the substances requiring the following R-phrases both the name of the corresponding substance(s) and the R-phrase should be mentioned in the label: R39, R40, R42, R43, R42/43, R45, R46, R47, R48, R49, R60, R61, R62, R63, R64. As a general rule a maximum of four R-phrases and four S-phrases should suffice to describe the risks and to formulate the most appropriate safety advice.

Symbols showing the most serious hazards should be chosen where more than one danger symbol has to be assigned. As a general rule a maximum of two danger symbols are used.

Table 1 explains the letter symbols appearing in the attached lists. Each letter symbol refers to a danger symbol or pictogram (Figure 6).

When more than one danger symbol is used:

- obligation to apply symbol T or T+ will make symbols C, Xn and Xi optional;
- obligation to apply symbol C will make symbols Xn and Xi optional;
- obligation to apply symbol E will make symbols F and O optional.

If a preparation is classified both harmful (Xn) and irritant (Xi), it will be labelled Xn, and the irritant properties should be pointed out with appropriate R-phrases. The total amount of the substance in the preparation must be considered in choosing the danger symbols, R- and S-phrases.

Generally, no account needs to be taken of substances if they are present in the following amounts, unless another lower limit has been specifically given:

less than 0.1% by weight for substances classified as very toxic T+, or toxic T
less than 1% for substances classified as harmful Xn, corrosive C, irritant Xi

Any pictorial symbol indicating danger is drawn in black and the background colour should be orange.

Table 2 gives the specified dimensions of the label.

The Material Safety Data Sheet (MSDS). Increasingly, there is also a requirement that manufacturers and suppliers provide a material safety data sheet, now often referred to only as a safety data sheet (SDS), with all the relevant hazard and safety information available. The term MSDS is slightly misleading since each MSDS may run to several A4 pages. The MSDS backs up the information that must be provided on the label attached to any container of a substance and is required by law in North America and the European Union. Similar laws are gradually being implemented worldwide as a result of the development of international agreement on a Globally Harmonized System (GHS) of chemicals management. MSDS must be kept on file, available to both workers and management, in the workplace at all times. However, as will be seen below, the usefulness of the MSDS is limited and its limitations must be appreciated if it is to be interpreted properly to guide safety precautions.

A standardized format is used in the preparation of an MSDS. Unfortunately, the information provided on some MSDSs is not validated or checked

Table 1 Letter symbols and their meaning.

Letter symbol	*Definition*	*Explanation*
E	Explosive	This symbol with the word 'explosive' denotes a substance that may explode under the effect of a flame or if subjected to shocks or friction
O	Oxidizing	The symbol with the word 'oxidizing' refers to a substance that releases excessive heat when it reacts with other substances, particularly with flammable substances
F	Highly flammable	This symbol with the words 'highly flammable' denotes a substance that may become hot and finally catch fire in contact with air at ambient temperature, or is a solid and may readily catch fire after brief contact with the source of ignition, and which continues to burn or to be consumed by chemical reaction after removal of the source of ignition. If it is gas it may burn in air at normal pressure. If it is a liquid it would catch fire with slight warming and exposure to a flame. In contact with water or damp air the substance may release highly flammable gases in dangerous quantities
F+	Extremely flammable	The same flammable symbol as above with words 'extremely flammable' denotes, for example, a liquid that would boil at body temperature and would catch fire if vapours are exposed to a flame
T	Toxic	The symbol with skull and crossed bones with the word 'toxic' denotes a highly hazardous substance
T+	Very toxic	The same symbol as above with the words 'very toxic' is used to label a substance that, if inhaled or ingested, or if it penetrates the skin, may involve extremely serious immediate or long-term health risks and even death
C	Corrosive	The symbol with the word 'corrosive' will be found on a label of a substance that may destroy living tissues on contact with them. Severe burns may result from splashes of such substance
Xn	Harmful (less than T)	The symbol with the word 'harmful' denotes substances that may cause health hazards less than toxic. It could refer to other types of risks, *e.g.* to allergic reactions
Xi	Irritant (less than C)	The same symbol as above with the word 'irritant'

Table 2 Specified label dimensions.[a]

Capacity of the package	*Minimum dimensions (mm)*
Not exceeding 3 L (litres)	52 × 74
More than 3 L but not exceeding 50 L	74 × 105
More than 50 L but not exceeding 500 L	105 × 148
More than 500 L	148 × 210

[a]*Note*: Each danger symbol must cover at least 1/10 of the surface area of the label. The minimum size of the danger symbol shall not be less than 10 × 10 mm.

and must be considered critically before any use is made of it. However, International Chemical Safety Cards, IPCS validated versions of MSDSs, are available on the International Programme on Chemical Safety (IPCS) Chemical Safety Information from Intergovernmental Organizations (INCHEM) web site with similar information for pure substances (<http://www.inchem.org>). In addition, there are also Concise International Chemical Assessment Documents (CICADS), which are short booklets produced by IPCS to summarize the key information required by industrial managements to protect their workforces. These documents also may be downloaded from the INCHEM web site. CICADS contain copies of the relevant International Chemical Safety Cards.

Example List of the Components of a Material Safety Data Sheet, With Annotations

Note: This list is based on United States regulations and other countries, such as those of the European Union, have similar but slightly different requirements. Guidance on the preparation of Safety Data Sheets for the Globally Harmonized System of Classification and Labeling of Chemicals (GHS) can be found at <http://www.unece.org/trans/danger/publi/ghs/ghs_rev02/02files_e.html>.</note>

Section I – Identity Information

Identity of the product: Identification of the substance by manufacturer, common name and synonyms, product code(s), chemical formula, chemical class classification (*e.g.* alcohol *etc.*), shipping name (for national and international recognition).

EMERGENCY TELEPHONE NUMBER: Must be included, but it does not need to be toll free.

TELEPHONE NUMBER FOR INFORMATION: May be the same as above for small companies.

NAME OF THE MANUFACTURER OR IMPORTER: Be sure that this name is exactly the same as the name of the manufacturer listed on the product label. Small manufacturers sometimes send out MSDS from the manufacturer of the raw materials they mixed to make the product or that they repackaged. This should not be the case.

ADDRESS OF THE MANUFACTURER: Be sure this address is complete: street or box, town, state and postcode (zip).

DATE PREPARED: MSDSs prepared more than three years ago are acceptable in the U.S., but an attempt should be made to get an updated version. Three-year-old MSDSs are invalid in Canada.

SIGNATURE OF THE PREPARER (OPTIONAL).

Section II – Hazardous Ingredients

List of ingredients with common names and synonyms, Chemical Abstracts Service (CAS) numbers, concentrations, legal exposure limits, American Conference of Government Industrial Hygienists (ACGIH) threshold limit values (TLVs) and source of information. Toxic chemicals comprising more than 1% of the product by weight must be listed. Cancer-causing chemicals that comprise 0.1% of the product must also be listed. If exposure to amounts even smaller than the required 1.0 or 0.1% is known to be hazardous, the manufacturer also must list these ingredients. In practice, however, such hazardous ingredients often go unlisted. For example, trace amounts of extremely toxic dioxins and PCBs in many pigments usually are not reported.

TRADE SECRET EXEMPTIONS: Information on the identity of hazardous ingredients can be withheld by the manufacturer if they are trade secrets. The MSDS should state by whose authority (usually the competent health department) the product's identity can be withheld. Trade secret products should be avoided whenever possible since it is very difficult and time-consuming for medical personnel to get this data if there is an accident or illness. Even then, the medical person must withhold from the victim the name of the chemical that caused his (her) problem.

Section III – Physical Data

Physical state and appearance; odour and odour threshold (level at which it would be detected in the nasal passages); the specific gravity; the vapour pressure (an index of the volatility at a specified temperature, usually at 20 °C); the vapour density – lighter or heavier than air; the evaporation rate relative to a reference chemical; boiling point; freezing point; the pH, if applicable (acidic, neutral or basic in nature – identifying corrosiveness); and, lastly, the oil/water partition coefficient (log K_{ow}).

> *Note*: Without knowing anything about the biological properties of the chemical, the physical properties permit the reader to make educated guesses concerning whether the agent will pose a health hazard by inhalation or by dermal contact, where the highest concentrations might be in a room following spillage (*e.g.* near the floor if it is much heavier than air), *etc.*

Section IV – Fire and Explosion Hazard

Mobility of the vapour; the flash point, flammable limits, auto-ignition temperature; explosion data – upper and lower explosive (flammable) limits (UEL, LEL) in air; percent by volume in air as well as mechanism(s) of initiating an explosion. In addition, extinguishing media are given as well as the potentially hazardous combustion products.

Note: This information is essential to fire fighting and rescue personnel.

Section V – Reactivity Data

Stability, incompatibility (materials to avoid), conditions to avoid, hazardous decomposition products, hazardous polymerization, *etc*.

Note: It may be necessary to obtain MSDSs for decomposition products if they are likely to be formed during storage or use of the primary material.

Section VI – Health Hazard Data

Primary route of exposure (dermal, oral, ocular, inhalation), over-exposure effects (irritation, sensitization) for acute and chronic effects, special toxicity (mutagenicity) teratogenicity, carcinogenicity, reproductive/fertility. Emergency and first aid procedures.

Note 1: Effects may differ widely depending on the route of exposure. Often the respiratory route is the most harmful but certain substances such as phenol penetrate skin rapidly. In the workplace, oral exposure is often neglected but carelessness can make it a significant source of harm.
Note 2: Sensitization leading to dermatitis or respiratory problems may result from low levels of exposure that do not themselves cause obvious immediate toxicity. Sensitization effects to humans may not be identified in animal studies and should always be considered possible and guarded against.
Note 3: Mutagenicity, teratogenicity, carcinogenicity and reproductive toxicity require extreme caution in handling substances that have such properties. If possible, such substances should not be used and suitable substitutes identified. This requires careful risk assessment, as poorly characterized substitutes may appear safer because of ignorance of their potentially harmful properties. Further, different risks will require some comparative evaluation. Thus, a low potency carcinogen that is an effective fire retardant may be preferable to a non-carcinogenic alternative that is less effective in preventing fires.

Section VII – Precautions for Safe Handling and Use

STEPS TO BE TAKEN IF MATERIAL IS RELEASED OR SPILLED: The MSDS should list preferred methods for spill control (*e.g.* chemical sorbents, Fuller's earth, *etc.*) and protective equipment (respirators, gloves, emergency ventilation, *etc.*) needed to keep workers safe during clean up of large spills or accidents.

Note: Read carefully before use of any substance and be sure you can dispose of it safely and without harming your local environment.

WASTE DISPOSAL METHOD: Unless the material can be rendered completely innocuous, the MSDSs can only tell users to dispose of the material in accordance with the existing regulations. Disposal has become an extraordinarily complex problem and cannot be addressed in a few lines on an MSDS. Substances that pose severe environmental threats or whose release (spills) must be reported should be identified here.

Section VIII – Control Measures

RESPIRATORY PROTECTION (SPECIFIC TYPE): If needed during normal use, a good MSDS explains precisely what type of respirator is proper. Even the type of cartridge type for air purifying respirators should be specified.

VENTILATION: If needed during normal use, a good MSDS specifies the type of ventilation system that provides proper protection. This includes recommendations about the use of general (mechanical) ventilation, local exhaust (which captures the contaminants at their source) or any special ventilation system that might be needed.

PROTECTIVE GLOVES: Good MSDSs list the specific type of glove material needed (rubber, nitrile, *etc.*) and other glove attributes such as length and thickness. Workers should know that some solvents penetrate gloves without changing the glove's appearance. Often such solvents are perceived only as perspiration. Good MSDSs indicate which gloves will resist penetration by the product. When in doubt, contact the technical department of your glove supplier.

EYE PROTECTION: Good MSDSs list precisely what type of goggles or glasses are needed by their ANSI Z87.1 standard classification. The MSDS at least should indicate whether vented or unvented chemicals splash goggles, impact goggles or other specific types are needed.

OTHER PROTECTIVE CLOTHING OR EQUIPMENT: Aprons, boots, face shields or eye wash stations should be listed here if needed.

WORK/HYGIENIC PRACTICES: Practices such as proper daily clean up methods and equipment after normal use should be detailed here.

Note: Make sure that you are prepared for any emergency before you use a chemical. Have all the appropriate equipment and any neutralizing materials readily available.

Section IX – Special Precautions

Handling, shipping and storage precautions, special warnings; regulatory information – national and (or) international; appropriate telephone numbers to contact expertise in cases of difficulties with a product.

> *Note*: This information should be checked carefully, especially if your use of the chemical is out of the ordinary.

Section X – Regulatory Information

DOT class number, departmental response numbers, government numbers, *e.g.* USA-TSCA, USA-RCRA, CERCLA status.

> *Note*: These regulatory values are useful for tracking information in official databases.

Further Reading

Health and Safety Executive (UK), *List of Symbols, Abbreviations, Risk and Safety Phrases*, 2008. Available at <http://www.hse.gov.uk/chip/phrases.htm>.

International Labour Organization (ILO), *Identification, Classification and Labelling of Chemicals. International Occupational Safety and Health Centre (CIS)*, 2008. Available at <http://www.ilo.org/public/english/protection/safework/cis/products/safetytm/classify.htm>.

International Programme on Chemical Safety (IPCS), *INCHEM (Chemical Safety Information from Intergovernmental Organizations)*, 2008. Available at <http://www.inchem.org/>.

MSDSonline, *Where to Find an MSDS on the Web*, 2008. Available at <http://www.ilpi.com/msds/>.

UNECE, *Globally Harmonized System of Classification and Labelling of Chemicals (GHS)*, UNECE, Geneva, 2nd revised edn, 2007. Available at <http://www.unece.org/trans/danger/publi/ghs/ghs_rev02/02files_e.html>.

1.6 Terms Applied to Toxic Substances

biocide
Substance intended to kill living organisms.

This term as defined above includes all pesticides and related substances. Somewhat confusingly, in European legislation the term 'biocides' excludes agricultural pesticides, plant protection products, medicines and cosmetics, which are covered by other more specific legislation, and includes only those pesticides that are used

Table 3 Groups of biocides covered by European legislation.

Main Group 1	*Disinfectants and General Biocidal Products*
Product-type 1	Human hygiene biocidal products
Product-type 2	Private area and public health area disinfectants and other biocidal products
Product-type 3	Veterinary hygiene biocidal products
Product-type 4	Food and feed area disinfectants
Product-type 5	Drinking water disinfectants
Main Group 2	*Preservatives*
Product-type 6	In-can preservatives
Product-type 7	Film preservatives
Product-type 8	Wood preservatives
Product-type 9	Fibre, leather, rubber and polymerized materials preservatives
Product-type 10	Masonry preservatives
Product-type 11	Preservatives for liquid-cooling and processing systems
Product-type 12	Slimicides
Product-type 13	Metalworking-fluid preservatives
Main Group 3	*Pest Control*
Product-type 14	Rodenticides
Product-type 15	Avicides
Product-type 16	Molluscicides
Product-type 17	Piscicides
Product-type 18	Insecticides, acaricides and products to control other arthropods
Product-type 19	Repellents and attractants
Main Group 4	*Other Biocidal Products*
Product-type 20	Preservatives for food or feedstocks
Product-type 21	Antifouling products
Product-type 22	Embalming and taxidermist fluids
Product-type 23	Control of other vertebrates

for certain restricted purposes such as preserving wood, preventing ship fouling, disinfection, controlling mice and rats, and controlling domestic insects such as cockroaches and ants. However, this still leaves in the European legislation a large number of potentially harmful substances of the types listed in Table 3.

drug
Any substance that, when absorbed into a living organism, may modify one or more of its functions.

The term is generally accepted for a substance taken for a therapeutic purpose, but is also commonly used for substances of abuse. Just as any substance can be a toxicant, so any substance can be a drug. The term carries with it the implication of use for medical purposes, but also the potential for abuse to produce an effect desired by the abuser, but which is ultimately harmful.

pesticide
Substance intended to kill pests: in common usage, any substance used for controlling, preventing, or destroying animal, microbiological, or plant pests.

As with most of the terms in this group, the definition can only be applied with knowledge of the intended use of the substance. Almost any substance can be a pesticide if it is used for that purpose. For example, acetone can be used to kill most insects, but it is unusual to use it for this purpose and so it is not normally classified as a pesticide. Similarly, sodium chloride can kill most weeds, but is not classified as a pesticide. Thus, the term as used in practice is usually based on some official list produced for regulatory purposes and has little scientific logic behind it. This may lead to careless use by people who do not understand that many pesticides are not specific for the pests to which they are applied but can harm people as well. This may be a particular problem with herbicides, which are named as though they were specific for killing plants although they may be – like paraquat – extremely toxic to humans.

poison (in toxicology)
Substance that, taken into or formed within the organism, impairs the health of the organism and may kill it.

This word comes from the Greek *potein* to drink and hence has the same root as the word 'potion'. When love potions were devised, their use to affect other people, often to their harm, gradually led to the idea of poison as we use the word today (see 'venom' below). It may be related to the Irish word 'poteen', which means illegally distilled Irish whisky.

toxicant
This is the preferred term for a substance that is considered to be toxic under circumstances which are thought likely to happen.

This word comes from the Greek *toxikos* = of or for the bow, and was originally applied to the poison used to tip arrows. The term 'poison' is nearly a synonym, but tends to be applied to substances that may be deliberately used for poisoning, such as pesticides, and often has overtones of criminal use.

toxic substance (agent, chemical, material)
Material causing injury to living organisms as a result of physicochemical interactions.

To the toxicologist, any substance is potentially toxic since it is a matter of dose and so the distinction between toxic and non-toxic is arbitrary. For regulatory purposes, it has been historically based on the short-term (acute) LD_{50}, but this is a very unsatisfactory basis for such classification as it is essentially a rather poor (single point) measure of the capacity to kill mammals as surrogates for humans and it is not an absolute measure since, even with similar test populations, it can vary considerably. Many animals have died for this classification, which tells us little about doses causing sublethal or chronic effects, for example, mutations or cancer, which are of major concern for human health. Thus, labelling substances as harmful, toxic or very toxic on the basis of the LD_{50} is of limited value as substances that cause serious sublethal and/or chronic effects may not be labelled as harmful or toxic despite their significant potential for harm.

toxin
Poisonous substance produced by a biological organism such as a microbe, animal, or plant.

Like 'toxicant', this word comes from the Greek *toxikos* = of or for the bow, and was originally applied to the poison, usually extracted from plants, that was used to tip the arrows. In turn, it may derive from *taxus*, the yew tree, from which arrows were made and which has berries that are poisonous. 'Toxin' has been commonly used as a synonym for 'toxicant', but this usage is unacceptable since the distinction between naturally occurring toxicants produced by living organisms (true 'toxins') and synthetic toxicants is an important one. It is particularly inappropriate to apply the term to inorganic toxicants since no inorganic toxin is known. Examples of true toxins are those such as botulinum toxin, tetrodotoxin, pyrrolizidine alkaloids or amanitin.

venom
Animal toxin generally used for self-defence or predation and usually delivered by a bite or sting.

This term derives from the word 'wen' meaning to wish, from which developed 'venus', 'venery' and venerate', all related to concepts of love. A love potion became a 'venin', and this became 'venom'.

xenobiotic
Compound with a chemical structure foreign to a given organism.

The term is usually restricted to manmade compounds. It originates from the Greek words *xenos* = foreign and *biotikos* = living. True toxins as defined above are never referred to as xenobiotics although they may occur in circumstances that satisfy the above definition.

Related Definitions

medicine
Any drug or remedy.

Again this definition depends on usage. Any substance, *e.g.* herbs, willow bark or honey, may be used as a drug or a remedy and, as always, the end effect will depend on the dose.

pharmaceutical
Medicinal drug.

The Greek root *pharmakon* also means 'enchanted potion' or 'poison'. This term may have a legal definition in national legislation.

Further Reading

European Union, DIRECTIVE 98/8/EC OF THE EUROPEAN PARLIAMENT AND OF THE COUNCIL of 16 February 1998 concerning the placing of biocidal products on the market, *Official J. Eur. Communities*, L 123/1, 1998. <ec.europa.eu/environment/biocides/pdf/dir_98_8_biocides.pdf>

UNECE, *Globally Harmonized System of Classification and Labelling of Chemicals (GHS)*, UNECE, Geneva, 2008: <http://www.unece.org/trans/danger/publi/ghs/ghs_rev02/02files_e.html>.

C. Winder, R. Azzi and D. Wagner, The development of the globally harmonized system (GHS) of classification and labelling of hazardous chemicals, *J. Hazard. Mater.*, 2005, **A125**, 29.

Subgroup B – Hazard and Risk Assessment

1.7 Hazard

hazard
Set of inherent properties of a substance, mixture of substances, or a process involving substances that, under production, usage or disposal conditions, make it capable of causing adverse effects to organisms or the environment, depending on the degree of exposure; in other words, it is a source of danger.

The IUPAC definition may not make it clear that the term 'hazard' may also be applied directly to the substance, agent, source of energy or situation having hazardous properties.

Just because a substance has hazardous properties does not mean that these properties will necessarily be expressed. For most toxic effects, substances must be present in relevant media at a concentration above a threshold level before any toxicity will be apparent as a result of the threshold dose being exceeded. For mutagenic, carcinogenic and teratogenic effects, for which it is assumed that there is no threshold, the hazardous properties must be considered in terms of risk. This requires determining the relationship of risk of these properties being expressed to exposure in terms of concentration and/or dose. This process will be discussed under the definition of 'risk'.

Hazard Assessment. Hazard assessment is the process designed to determine factors contributing to the possible adverse effects of a substance to which a human population or an environmental compartment could be exposed. The process includes three steps: hazard identification, hazard characterization and hazard evaluation (Figure 7). Factors affecting toxicity may include metabolism, dose–effect and dose–response relationships, and variations in target susceptibility, amongst others.

Hazard Identification. The first stage in hazard assessment is the determination of substances of concern and the adverse effects that they may have on target systems under defined conditions of exposure, taking into account all the data relating to toxicity, especially relevant physicochemical properties such as volatility and solubility and aerodynamic diameter, particle size. A list is made of these substances along all the available relevant

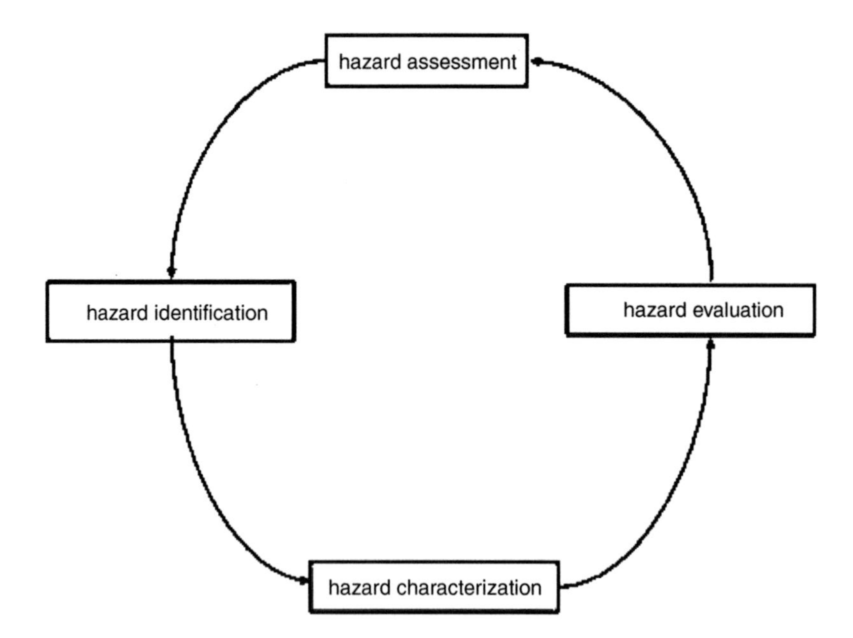

Figure 7 Hazard assessment concept diagram.

information. Usually the relevant information will be available in the form of material safety data sheets (MSDS). In many countries there is a legal requirement that MSDS must be provided by chemical producers or suppliers. It is also a normal legal requirement that chemicals should be labelled in such a manner that hazardous properties are clearly identified. In the European Union, appropriate 'risk' and 'safety' phrases must be included on the label in the language of the country where the chemicals are to be used. These phrases are allocated by an international expert group involving representation from the International Programme on Chemical Safety (IPCS). The phrases are also to be found with other evaluated hazard information on the International Chemical Safety Cards produced by IPCS and available on the INCHEM web site and the US National Institute of Occupational Safety and Health (NIOSH) web site.

Hazard Characterization. This is the second step in the process of hazard assessment. It consists of the qualitative and, wherever possible, quantitative description of the nature of the hazard associated with the agent of concern. The description may cover various aspects of the hazard considered in a holistic way. Thus, attention is given to the mechanisms of action of the agent, the biological extrapolation of these mechanisms to physiological consequences and dose–response and dose–effect relationships, amongst other properties likely to be relevant to the specific circumstances under consideration. It is particularly important to define as precisely as possible any related uncertainties as these are essential for subsequent risk assessment.

Hazard Evaluation. This is the third step in the process of hazard assessment aiming at the determination of the qualitative and quantitative relationship between exposure to a hazard and the resultant adverse effects under the defined conditions of exposure that may give concern. As in hazard characterization, it is important to define and include any attendant uncertainties. In risk assessment, assessment of probability of exposure is related to hazard evaluation and a level of exposure likely to produce a significant level of harm (Section 1.2.3).

Hazard Quotient (HQ). The ratio of toxicant exposure (estimated or measured) to a reference value regarded as corresponding to a threshold of toxicity: if the total hazard quotient from all toxicants to a target exceeds unity, the combination of toxicants may produce (will produce under assumptions of additivity) an adverse effect.

Further Reading

P. Illing, *Toxicity and Risk: Context, Principles and Practice*, Taylor and Francis, London, CRC Press, Boca Raton, FL, 2001.

1.8 Structure–Activity Relationship (SAR)

structure–activity relationship (SAR)
Association between specific aspects of molecular structure and defined biological action. See also quantitative structure–activity relationship.
structure–metabolism relationship (SMR)
Association between the physicochemical and/or the structural properties of a substance and its metabolic behaviour.
quantitative structure–activity relationship (QSAR)
Quantitative structure–biological activity models derived using regression analysis and containing as parameters physicochemical constants, indicator variables, or theoretically calculated values.

Note: The term is extended by some authors to include chemical reactivity, where activity and reactivity are regarded as synonyms. This extension is deprecated.

quantitative structure–metabolism relationship (QSMR)
Quantitative association between the physicochemical and (or) the structural properties of a substance and its metabolic behaviour.

In toxicology, SAR methods apply various mathematical and statistical models to predict the adverse effects of chemicals based upon their structure. The prediction may be qualitative (*e.g.* is a substance likely to cause cancer?) or quantitative, QSAR (what level of dose will produce a given effect?). Such methods give results that can be used in various ways. They may indicate a need for further experimentation and evaluation and can be used to select toxicity tests for predicted end-points of concern. This includes prioritizing tests so that likely effects are tested first, which may eliminate the need for further testing.

There is a hope that SAR methods will eventually be an adequate replacement for animal testing, but the current state of the art is not good enough to permit this. In particular, the more possible mechanisms that are associated with an effect, the more difficult and consequently less accurate is any prediction. However, in drug development, animal testing may be avoided for certain compounds for which a SAR clearly indicates the potential for serious adverse effects.

Elucidation of SARs is best developed for organic compounds and is still poorly developed for inorganic compounds. For organic compounds, identification of SARs requires knowledge of the biological activities of defined chemical structures, of biological interactions when structures occur in the same molecule, and the derivation of models that can be used to relate total molecular structure to biological effects.

Creation of models uses physicochemical data along with manual pattern recognition methods, cluster analysis and regression analysis. For meaningful models, data from a substantial number of compounds with differing substituent combinations and well-defined biological effects are required. The main difficulty is in analyzing the data to identify particular structural fragments responsible for the production of a defined effect. Even if such fragments are identified, the question remains as to whether these fragments are sufficient in themselves to produce the effect, whether they are always necessary for this effect and whether the effect is modified by the molecular environment.

SAR methods are available for organic molecules to predict genotoxicity, carcinogenesis, dermal irritation and sensitisation, lethality, biological oxygen demand and teratogenicity, with varying degrees of accuracy. Both the USEPA (United States Environment Protection Agency) and USFDA (United States Food and Drug Agency) use models for mutagenicity and carcinogenicity to screen for possible problem compounds.

QSAR. The common view of toxicologists is that QSAR is a screening tool that will catch approximately 60–70% of the tested end-point. QSAR has been validated in several studies that show that two-thirds of expected toxicity will be predicted. The corresponding figure for genotoxicity and carcinogenicity can be more than 70%. This has been found for a restricted number of chemicals. For reproductive toxicity, the predictive value is low. It is important to keep in mind that QSAR is based on available databases, which means that there is a limited knowledge with regard to new chemicals and also to chemicals with presently uncharacterized effects. These attempts at quantitative predictions are currently imprecise and inaccurate, but the field is developing rapidly and QSAR is a valuable tool when used with caution.

Further Reading

C. A. Lipinski, F. Lombardo, B. W. Dominy and P. J. Feeney, Experimental and computational approaches to estimate solubility and permeability in drug discovery and development settings, *Adv. Drug Delivery Rev.*, 2001, **46**, 3.

1.9 Risk

risk
1 Probability of adverse effects caused under specified circumstances by an agent in an organism, a population, or an ecological system.
2 Expected frequency of occurrence of a harmful event arising from such an exposure.

risk assessment
Identification and quantification of the risk resulting from a specific use or occurrence of a chemical or physical agent, taking into account possible harmful effects on individuals or populations exposed to the agent in the amount and manner proposed and all the possible routes of exposure.

Note: Quantification ideally requires the establishment of dose–effect and dose–response relationships in likely target individuals and populations.

Emphasis is placed on the concept of risk as a measure of probability. There is no mention here of the severity of the adverse effects that is sometimes incorporated in definitions of risk such as risk = (probability of unwanted event) × (severity of event). This is because assessment of severity, except at extremes, is essentially a subjective judgment and is part of the definition of hazard. It is important to keep considerations of risk as objective as possible because they determine what management decisions are to be taken following risk assessment (see below). If management decisions are to be effective, they must be accepted by those to whom they apply. Acceptance depends first of all on agreement on the level of risk. It is therefore important to eliminate subjective elements as far as possible from this first stage of risk assessment. The second stage involves the subjective determination of risk acceptability (see below) and this depends, amongst other things, on perception of the severity of the adverse effect for which the risk has been determined.

Risk Assessment. This is the identification and quantification of the risk resulting from a specific use or occurrence of an agent, taking into account possible harmful effects on individuals exposed to the agent in the amount and manner proposed and all the possible routes of exposure. Quantification requires the establishment of dose–effect and dose–response relationships in likely target individuals and populations. The process includes four steps (Figure 8): hazard identification, dose–response assessment, exposure assessment and risk characterization.

Hazard Identification. This has already described under 'hazard' above, but the other concepts in the diagram are defined as below.

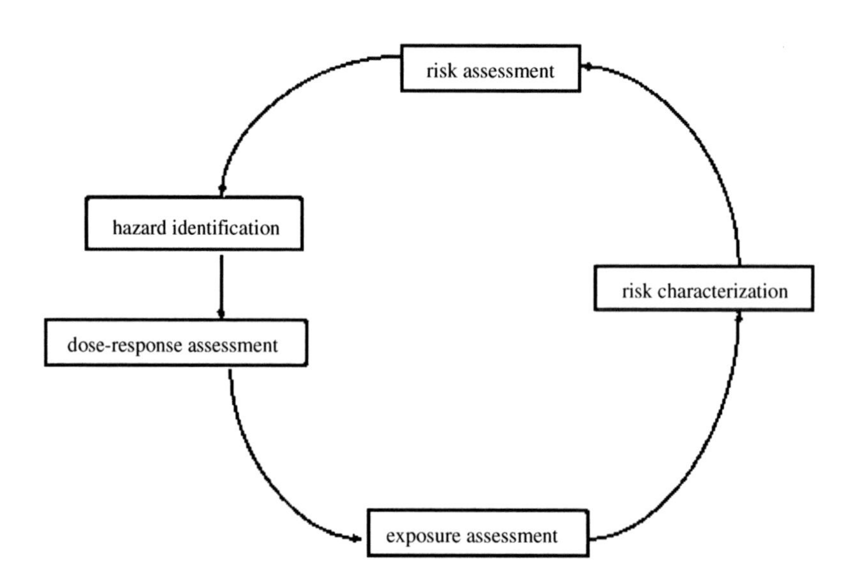

Figure 8 Risk assessment concept diagram.

Dose–Response Assessment. The second of four steps in risk assessment, consisting of the analysis of the relationship between the total amount of an agent absorbed by each of a group of organisms and the changes developed in the group in reaction to the agent, and inferences derived from such an analysis with respect to the entire population.

Notably, dose–response assessment always involves extrapolation of results from an experimental or observational group (a sample) to an entire population. Thus, there is a degree of uncertainty resulting from this procedure. Such uncertainty should be determined statistically and clearly stated to inform subsequent management decisions.

Exposure Assessment. Third step in the process of risk assessment:

1. Quantitative and qualitative analysis of the presence of an agent (including its derivatives) that may be present in a given environment and the inference of the possible consequences it may have for a given population of particular concern. Exposure often changes with time and so a calculated mean exposure over a given time interval may be used as the basis for risk assessment. In this case, special attention may need to be given to the possible occurrence of short-term peaks of extreme exposure.
2. Determination, using a range of different techniques, of the amount of a chemical, physical or biological agent that could be present in a given medium and the fate of such agent under several potential circumstances, and the inference of possible consequences for a hypothetical system that could be affected by this agent. This determination may be based on theoretical considerations and use computer modelling to predict movement and distribution of a substance within one or more environmental compartments at risk.

Risk Characterization. *Preferred Definition.* Integration of evidence, reasoning and conclusions collected in hazard identification, dose–response assessment and exposure assessment and the estimation of the probability, including attendant uncertainties, of occurrence of an adverse effect if an agent is administered to, taken by or absorbed by a particular organism or population. It is the last step of risk assessment.

In ecological risk assessment, concentration–response assessment is carried out instead of dose–response assessment.

Alternative Definition. Qualitative and (or) quantitative estimation, including attendant uncertainties, of the severity and probability of occurrence of known and potential adverse effects of a substance in a given population.

This definition requires consideration of all possible effects and their severity. Although this is a good objective, it is unlikely to be attainable in practice because of its complexity. In the end, the objective can be most nearly reached by combining risk characterizations carried out according to the preferred definition.

Risk Evaluation. This involves the establishment of a qualitative or quantitative relationship between risks and benefits, involving the complex process of determining the significance of the identified hazards and estimated risks to those organisms or people concerned with or affected by them. It is the first step in risk management and includes economic, ethical and other non-scientific considerations. "Risk evaluation" is synonymous for "risk–benefit evaluation".

To compare risk and benefit, these concepts must be defined in a compatible manner. Since risk is defined in terms of probability of harm, benefit should also be defined in terms of the probability of a good outcome. It will also help if 'harm' and 'good outcome' can be quantitatively defined in the same units.

Emission and Exposure Control. This is the regulation of emission of potentially toxic substances to ensure that exposure to these substances is kept below a level likely to be harmful to humans or other species at risk.

This usually involves establishing and agreeing guidelines or legal standards for acceptable ambient concentrations in environmental media together with a system for monitoring exposure and enforcement of standards.

Risk Monitoring. Risk monitoring is the process of following up the decisions and actions within risk management to ascertain that risk containment or reduction with respect to a particular hazard is assured.

Acceptable Risk. This is the type of risk such that the perceived benefits derived by an organism, a population or an ecological system outweigh the adverse effects that might affect them as a result of administration or exposure to a particular agent.

Acceptance of risk is subjective and dependent upon perception. Different people and groups of people may have very different perceptions, and thus 'acceptable risk' can have no absolute definition, for example, in terms of a certain level of probability.

Risk Management. Risk management is the decision-making process involving considerations of political, social, economic and technical factors with relevant risk assessment information relating to a hazard so as to develop, analyse and compare regulatory and non-regulatory options and to select and implement the optimal decisions and actions for safety from that hazard. Essentially, risk management is the combination of three steps: risk evaluation, emission and exposure control, and risk monitoring. These steps have been defined above.

Further Reading

P. Illing, *Toxicity and Risk: Context, Principles and Practice*, Taylor and Francis, London, CRC Press, Boca Raton, FL, 2001.

1.10 Risk Assessment, Risk Management and Safety

risk assessment
Identification and quantification of the risk resulting from a specific use or occurrence of a chemical or physical agent, taking into account possible harmful effects on individuals or populations exposed to the agent in the amount and manner proposed and all the possible routes of exposure.
Note: Quantification ideally requires the establishment of dose–effect and dose–response relationships in likely target individuals and populations.

risk assessment management process
Global term for the whole process from hazard identification to risk management.

risk management
Decision-making process involving considerations of political, social, economic, and engineering factors with relevant risk assessments relating to a potential hazard so as to develop, analyse, and compare regulatory options and to select the optimal regulatory response for safety from that hazard.
Note: Essentially risk management is the combination of three steps: risk evaluation, emission and exposure control and risk monitoring (as described above).

safety
Reciprocal of risk: practical certainty that injury will not result from a hazard under defined conditions.
Note 1: Safety of a drug or other substance in the context of human health: the extent to which a substance may be used in the amount necessary for the intended therapeutic purpose with a minimum risk of adverse health effects.
Note 2: Safety (toxicological): The high probability that injury will not result from exposure to a substance under defined conditions of quantity and manner of use, ideally controlled to minimize exposure.

Risk assessment, risk management and risk communication, combined as a process for controlling situations where an organism, system or (sub)population could be exposed to a hazard, has been defined as 'risk analysis' by IPCS and OECD (*see* Further Reading), following the precedent of the Joint Expert Committee on Food Additives (JECFA) of the World Health Organization (WHO). This use of the term 'analysis' must be deprecated as the process is actually one of integration and not analysis. By any conventional usage, 'analysis' refers to 'the resolution or breaking up of anything complex into its various simple elements', as defined in the current *Oxford English Dictionary*. There is no precedent for the perverse usage in the OECD/JECFA definition, which tends to confuse rather than to clarify thought.

The concept of risk has already been discussed above. Here the concern is with the provision of safety, *i.e.* the reduction of risk to an acceptable level.

'Acceptability' of risk varies from person to person and is, therefore, subjective, but risk assessment, as defined, should be objective. Thereafter, risk management should aim to reduce risk to levels acceptable to those at risk. This raises issues of risk communication that are outside the remit of this explanatory book since they involve consideration of public relations and the related psychological issues that vary as much as the groups at risk. Each group at risk has a different perception of risk, often dependent upon deeply rooted fundamental ideas, which are not open to debate, and must be accommodated for effective communication of risk and how it can be minimized.

Since all substances are toxic at a high enough dose, there is none for which one can guarantee absolute chemical safety, *i.e.* no risk of harm. Thus, the aim of chemical safety management is to reduce the risk associated with defined chemical processes to an acceptable minimum. This requires consideration of intrinsic hazard and the probability of exposure at a level that may be high enough to cause harm under defined conditions; in other words, risk assessment. Risk assessment is followed by management of the risk situation to minimize risk and therefore to optimize safety.

Risk to be considered for risk management purposes should include known quantifiable risks, known, but not quantifiable risks (which may be identifiable by testing), and unknown risks (that are identifiable only once the chemical is in use, *i.e.* through risk monitoring). Risk monitoring is an essential part of risk management. One example of risk monitoring is pharmacovigilance. Another is observational environmental monitoring.

Risk Assessment. There are four steps in risk assessment:

1. Hazard identification. Identify agents responsible for potential harmful effects, collect information on their physicochemical and harmful properties, on the group(s) of people or other living organisms liable to exposure, and on the exposure circumstances (*e.g.* environmental conditions).
2. Risk characterization. Identify potential effects of possible exposures and quantify dose–effect and dose–response relationships.
3. Exposure assessment. Quantify exposure, by measurement or biological monitoring if it is already occurring or by predictive modelling if it is not.
4. Risk estimation. Quantify risk in the exposed group, making clear any assumptions involved. Ideally, this should give a statistically based measure of probability of harm to the defined group under the defined circumstances. Often this is not possible and a *risk quotient* is calculated that is related to probability of harm but only as a rough approximation.

Risk Management. Following risk assessment, risk management is the process of deciding how to reduce to a practicable minimum the risk assessed. Again, various stages have been defined for this process:

1. Risk evaluation. Comparison of risks resulting from different actions and their probable consequences. This includes comparison of costs and

benefits. These comparisons are used to decide what might be an acceptable risk level, bearing in mind any irreducible level of risk that may already exist.

2. Exposure control. Actions required to keep exposure below the level associated with unacceptable risk. Normally this will be an exposure level that causes no risk, but for mutagens and carcinogens a risk of one in a million is generally regarded as acceptable.
3. Risk monitoring. Observation, assessment and measurement to determine how effective the actions taken to control exposure have been. This may include biomonitoring.
4. Risk management evaluation and improved exposure control. There is always an element of risk associated with failure of control for which accident (emergency) planning should be in place. If previous actions have not been as effective as expected, further actions must be taken to reduce exposure. Then stage 3 is repeated and so on until exposure control, and thus safety, is acceptable. Risk management is concerned with the consequences of risk evaluation. If risks are evaluated as broadly acceptable, no further management action may be needed. However, management may be required to reduce tolerable risks to a level at which they are broadly acceptable. Even if it is not possible to do this, risks should always be reduced as far as is reasonably practicable.

Risk management is a combination of prevention (or minimization) of exposure to a potentially harmful amount or concentration of a chemical and prevention (or minimization) of any consequences (ill-health and or pollution). The objectives are often achieved by one or more of the following approaches:

(a) Substitution [favoured in the European Registration, Evaluation, Authorisation and Restriction of Chemicals (REACH) Regulation]. Substitution of a substance by a less hazardous alternative or use of an alternative, less risky process would seem to be an obvious way to reduce risk, but such a change may reduce one risk at the cost of increasing another. For example, chlorofluorocarbons, which attack the ozone layer but are of low toxicity to humans, were replaced as a refrigerant by ammonia, which is highly toxic to humans and has injured people following accidental releases. DDT, which is of low toxicity to humans but poisons birds of prey after biomagnification, was replaced as an insecticide by organophosphates, which are highly toxic to humans. Now, DDT is being used again, with more care than before, since it is the most effective insecticide against *Anopheles* mosquitoes that were spreading malaria killing millions of people. Fire retardants with a potential for human toxicity may be substituted by less effective fire retardants because of fears of human toxicity and this may result in more fires, with accompanying mortality. These are a few examples of problems in substituting one substance by another that is apparently less hazardous. Perhaps the biggest problem in looking for a substitute for a given substance is that substances for which toxicity data are not fully available will always appear to be safer than those that have been thoroughly investigated.

Often, assurance that there is 'no evidence of toxicity' simply means that available data are minimal and approximate to 'no evidence at all' on which to make an informed judgment. It should be remembered that 'absence of evidence' is not 'evidence of absence'.

(b) Change of process. Changing the process may prevent or reduce exposure and (or) minimize emissions. Change may involve improving process containment or improving ventilation, and (or) secondary containment, *e.g.* collecting a substance before it is vented or discharged, perhaps by means of suitable filters, followed by reclamation and (or) recycling. In some circumstances, dilution may permit hazardous waste material to be safely discharged to an appropriate receiving environment where it can be degraded without harm.

(c) Use of personal protective equipment. This is the least favoured approach to reducing risk. Ideally, any exposure occurring during the use of a substance should be lower than that which can harm humans or their environment. Any process that requires wearing protective equipment, clothing, gloves or footwear is intrinsically of high risk. Care must always be taken to ensure that the equipment, clothing, *etc.* are appropriate and functioning properly.

As a general principle substitution is preferred to change of process, which is preferred to use of protective equipment. Measurements of concentrations of substances in the immediate environment, where the chemical is used, and health and ecological monitoring are necessary to ensure that controls are adequate. Biomonitoring may also be used, but invasive biomonitoring is generally not welcomed by people, and may itself cause undue stress on workers.

Risk management must include measures to ensure safety (and pollution) awareness amongst those at risk, especially members of the general public. Such awareness is dependent on public perceptions, as already mentioned. There are prejudices to be overcome, *e.g.* that manmade chemicals are more toxic than those that occur naturally, or that one cannot have too much of a substance that is essential to health such as a vitamin, or even water. Perhaps the greatest prejudice is that hazardous substances should be avoided on the basis of the 'precautionary principle'. Since every substance is hazardous at some exposure level, this prejudice is, in essence, suicidal. Thus, risk management demands public education to ensure that every person at risk understands the rationale of regulatory requirements and thus implements them in an appropriate manner.

Further Reading

EC (2004), *European Union System for the Evaluation of Substances 2.0 (EUSES 2.0)*, Prepared for the European Chemicals Bureau by the National Institute of Public Health and the Environment (RIVM),

Bilthoven, The Netherlands (RIVM Report no. 601900005). Available *via* the European Chemicals Bureau at <http://ecb.jrc.it>.

International Programme on Chemical Safety (IPCS), *Principles for the Assessment of Risks to Human Health from Exposure to Chemicals, Environmental Health Criteria 210*, WHO, Geneva (1999). Available from the IPCS-INCHEM website at <http://www.inchem.org/>.

International Programme on Chemical Safety (IPCS), *IPCS Risk Assessment Terminology, Part 1: IPCS/OECD Key Generic Terms used in Chemical Hazard/Risk Assessment: Part 2: IPCS Glossary of Key Exposure Assessment Terminology: Harmonization Project Document No. 1*, WHO, Geneva, 2004. Available at: <www.inchem.org/documents/harmproj/harmproj/harmproj1.pdf>.

Concept Group 2. Concepts Applying to Molecular and Cellular Toxicology

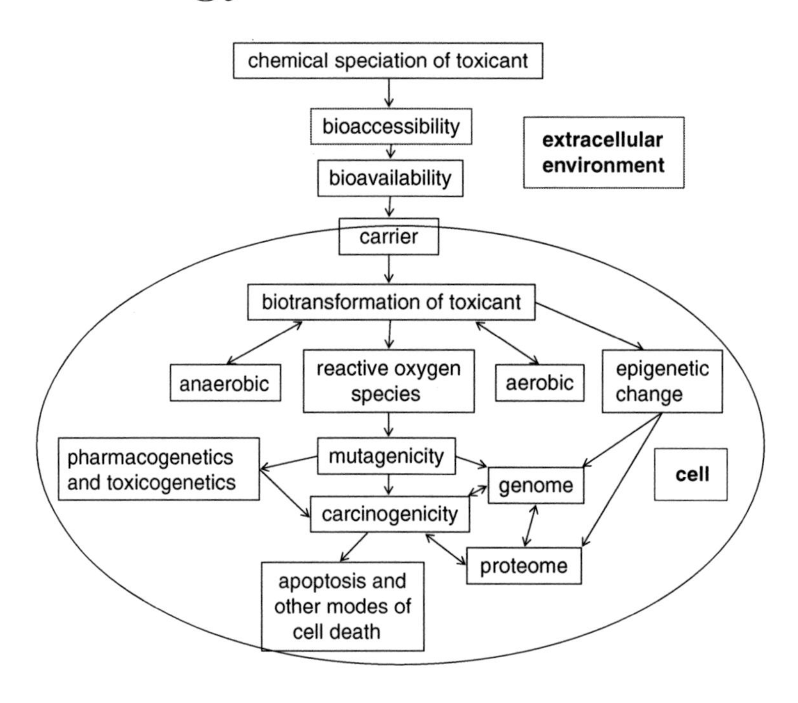

Concepts in Toxicology
By John H Duffus, Douglas M Templeton and Monica Nordberg

Published by the Royal Society of Chemistry, www.rsc.org

2.1 Speciation: Chemical and Biological

speciation (in biology)
1 Systematic classification of groups of organisms of common ancestry that are able to reproduce only among themselves, and that are usually geographically distinct.
2 Segregation of living organisms into groups that are able to reproduce only among themselves, and that are usually geographically distinct.
speciation (in chemistry)
Distribution of an element amongst defined chemical species in a system.
species (in biology)
In biological systematics, group of organisms of common ancestry that are able to reproduce only among themselves, and that are usually geographically distinct.
species (in chemistry)
Specific form of an element defined as to isotopic composition, electronic or oxidation state, and/or complex or molecular structure.

Speciation has two distinct meanings, one in chemistry and one in biology. Chemists speak of distinct species of an element, referring to different chemical forms, whereas biologists use the term to describe the origin or existence of groups of organisms that are closely related genetically. IUPAC has provided a definition of chemical species, and also recognizes the meaning of species in biology. Although not specifically a concept in molecular and cellular toxicology, biological speciation is considered here to distinguish clearly the two important uses of the terms.

Speciation in Chemistry. The terms 'species' and 'speciation' have become widely used in the chemical and toxicological literature, and it is now well established that the occurrence of an element in different compounds and forms is crucial to understanding the environmental and occupational toxicity of that element. Several definitions of chemical speciation can be found in the literature. In the past, the term 'speciation' has been used to refer to 'reaction specificity' (rarely); in geochemistry and environmental chemistry, to changes taking place during natural cycles of an element (species transformation); to the analytical activity of measuring the distribution of an element among species in a sample (speciation analysis); and to the distribution itself of an element among different species in a sample (species distribution). After a series of International Symposia on Speciation of Elements in Toxicology and in Environmental and Biological Sciences, the organizers formulated the definition 'Speciation is the occurrence of an element in separate, identifiable forms (*i.e.* chemical, physical or morphological state)'. The aim was to include determinants of reactivity, and produced a definition that goes well beyond speciation in a chemical sense and would include different phases of a pure substance, and even different-sized particles of a single compound. This selection of definitions illustrates the circularity in defining a species in terms of an entity, form or compound and indicates the lack of prior consensus in use of the term 'speciation'. To attempt to harmonize the field and offer at least partial solutions to the ambiguities present in some of the earlier definitions, IUPAC formulated and adopted the definitions cited above.

Fundamental to these concepts is the meaning of the term (chemical) 'species'. A chemical species is a specific form of an element that can be defined as distinct on several levels. Thus, at the lowest level two distinct species may differ only in their isotopic composition. In terms of human health and risk assessment, though, some structural aspects of speciation are more important than others. For instance, of much greater importance in toxicology than isotopic composition are differences in electronic or oxidation state. Also critical for understanding toxicity are differences in the molecular structures and complexes in which the element participates. Macromolecular species are excluded from the definition unless a macromolecular ligand is specifically defined. For example, a metal ion bound to two isoforms of a protein with defined amino acid sequences could be considered two species, but an ion bound to a polyelectrolyte such as humic acid or heparin would not be defined in terms of multiple species representing individual molecules in the heterogeneous and polydisperse population. In this case, it is advisable to refer to a fraction. 'Fractionation' can be defined as the process of classification of an element according to physical (*e.g.* size, solubility) or chemical (*e.g.* bonding, reactivity) properties. In this context, 'speciation' is defined as the distribution of an element among defined chemical 'species' in a system, and 'speciation analysis' is the analytical activity of identifying and (or) measuring the quantities of one or more individual chemical species in a sample.

Some examples of the importance to toxicology of speciation at the level of oxidation state, inorganic and organic complex formation, and organometallic compounds are illustrative. The greatly differing toxicities of Cr^{III} (non-toxic at usual exposures in the cation $[Cr(H_2O)_6OH]^{2+}$) and Cr^{VI} in the chromate anion $(Cr^{VI}O_4)^{3-}$ (a carcinogen) result because chromate anion $(Cr^{VI}O_4)^{3-}$ is taken up more readily by cells through anion transporters and subsequently Cr^{VI} releases electrons during its intracellular reduction to Cr^{III}. In contrast, As^{III} in arsenite is more toxic than As^V in arsenate, in part due to its greater ability to bind thiols in the lower oxidation state. Oxidation of Hg^0 vapour to Hg^{2+} by intracellular enzymes (*e.g.* in erythrocytes, neurons) causes the mercuric ion to become trapped in cells. Fe^{II} and Fe^{III} differ in their solubilities and have distinct transporter systems to get in and out of cells. Good examples of the importance of inorganic complexation are the various nickel compounds, which differ greatly in their carcinogenicity. For example, nickel disulfide (Ni_3S_2) is a potent carcinogen in rats but amorphous nickel sulfide (NiS) is not. Nickel salts are immune sensitizers, and the species of nickel can affect the physical form and thus the route of effective bioavailability. For example, nickel tetracarbonyl is a gas that is readily inhaled and absorbed from the lungs. Complexation of metals with organic ligands can affect their availability for cell uptake and for excretion from the body. Organometallic compounds in general are lipophilic and one consequence of this is an ability to penetrate the blood–brain barrier. Thus, whereas mercuric salts are peripheral neurotoxicants, alkylmercurials are potent central nervous system toxicants, as are many other alkylmetallics such as those of lead and tin.

Strictly speaking, whenever an element is present in different states according to isotopic composition, electronic or oxidation state, and (or) complex or molecular structure, it must be regarded as occurring in different species. In practice, however, usage will depend on the relevance of the species differences for our understanding of the system under study. One would not generally describe a living organism or define an organic reaction mechanism by carbon speciation. Nevertheless, a pair of stereoisomers are certainly distinct species, with different biochemical properties [*e.g.* only (*S*)-amino acids and (*R*)-sugars are used by living organisms], and if each formed a chelation complex with a metal ion these would be referred to as distinct species of the metal. Further, while the definition of species is general, in practice it is used mainly in the context of metallic and metalloid elements. Usage of speciation terminology also depends on our ability to distinguish the various species analytically. This practical analytical consideration governs whether different species should be grouped together or measured separately. Separate measurement implies minimum lifetimes and thermodynamic stabilities to allow detection, the values of which will change with developments in instrumentation.

Speciation in Biology. In the taxonomic subdivision of organisms, species is the lowest level of distinction. It is an uncapitalized subdivision of *genus*, and refers to organisms that are so closely related genetically that they successfully breed amongst themselves and generally not with other species; thus they carry on the genetic lineage. As the definition underscores, this close genetic similarity arises from common ancestry. Geographic isolation allows for the development of distinct species that subsequently become unable to interbreed with others. Indeed, the variety of unique species Charles Darwin encountered on his voyage to the geographically isolated Galapagos Islands was fundamental to the development of evolutionary theory.

In addition to the taxonomic usage of the above definitions, then, speciation also refers to the processes by which new species arise through evolution. While these now include deliberate laboratory manipulations and interventions such as animal husbandry and genetic modification of crops, natural speciation is worthy of discussion. There are generally considered to be four mechanisms of natural speciation, defined according to the degree and nature of geographic separation of the population undergoing speciation (the 'speciating population'): these are allopatric, peripatric, parapatric and sympatric:

1. Allopatric speciation occurs when a large population has been isolated by geographic barriers. If the barriers break down, allopatric species are found to have evolved so as to be unable to interbreed with other populations of what was originally the same species. They then constitute a new species.
2. Parapatric speciation refers to the evolution of two divergent species when geographic isolation is incomplete, there is some contact, but genetic separation still occurs because the heterozygote is not favoured by the prevailing environmental conditions.
3. Peripatric speciation occurs when small populations are isolated for various idiosyncratic reasons and stop breeding with the main population. They evolve in niches.
4. Sympatric speciation allows for species divergence without geographic separation. For example, groups of insects may start to feed on two different plant species growing in the same location, and eventually the different feeding groups become genetically distinct. A subtlety here is when different groups or races coexist but exploit different niches for behavioural reasons. This has been referred to as heteropatric speciation.

Two other terms should be mentioned. Cladogenesis refers to the branching of a new species when the founding one continues to exist. A clade is a group of organisms descending from a common ancestor. Anagenesis is the evolution of one species into another without divergence of the phylogenetic tree and with loss of the founder species.

While some see evolution as a gradual process, Stephen Jay Gould and others have proposed a theory of 'punctuated equilibrium' that proposes most species are stable throughout most of their geological history, and evolve in 'bursts'. Catastrophes and other stochastic events could account for these bursts, of which there is evidence in the palaeontological record. However, isolation of small groups from the large gene pool is viewed as a primary event in allowing rapid selection of favourable characteristics. The debate between 'punctuationalism' and 'gradualism' remains one of the more interesting discussions in the field of biological speciation.

Further Reading

D. M. Templeton, F. Ariese, R. Cornelis, L.-G. Danielsson, H. Muntau, H. P. van Leeuwen and R. Lobinski, Guidelines for terms related to chemical speciation and fractionation of elements. Definitions, structural aspects, and methodological approaches (IUPAC Recommendations 2000), *Pure Appl. Chem.*, 2000, **72**, 1453. Available at <http://www.iupac.org/publications/pac/72/8/>.

IPCS, *Elemental Speciation in Human Health Risk Assessment*, Environmental Health Criteria 234, WHO, Geneva, 2006. Available at International Programme on Chemical Safety (IPCS) INCHEM (Chemical Safety Information from Intergovernmental Organizations), 2008. Website <http://www.inchem.org/>.

2.2 Bioaccessibility and Bioavailability (See also Section 4.5)

bioaccessibility
Potential for a substance to come in contact with a living organism and then interact with it. This may lead to absorption.

Note: A substance trapped inside an insoluble particle is not bioaccessible although substances on the surface of the same particle are accessible and may also be bioavailable. Bioaccessibility, like bioavailability, is a function of both chemical speciation and biological properties. Even surface-bound substances may not be accessible to organisms that require the substances to be in solution.

bioavailability (general)
biological availability
physiological availability
Extent of absorption of a substance by a living organism compared to a standard system.

bioavailability (in toxico- or pharmacokinetics)
Ratio of the systemic exposure from extravascular (ev) exposure to that following intravenous (iv) exposure as described by the equation $F = A_{ev}D_{iv}/B_{iv}D_{ev}$, where F (fraction of dose absorbed) is a measure of the bioavailability, A and B are the areas under the (plasma) concentration–time curve following extravascular and intravenous administration, respectively, and D_{ev} and D_{iv} are the administered extravascular and intravenous doses.

Substances are biologically available if they can be taken up by living cells and organisms and can interact with 'target' molecules, including those on the cell surface. Thus, in the strictest sense of the term, it describes availability at the ultimate receptors. Measurement of the amount reaching the receptors is usually not possible. Hence, surrogate measurements are normally used. For humans, these may be levels found in blood or plasma. For plants, tissue concentrations may be used. For unicells such as protozoa and bacteria, the cell content may be appropriate.

Substances that are not bioavailable may still cause physical damage or may alter the bioavailability of other substances. The bioavailability of any element depends upon its chemical speciation (Section 2.1), but for many elements the determinants are poorly understood. Whereas bioavailability of many small organic molecules (*i.e.* carbon species) can often be predicted or explained on the basis of their physicochemical properties, the bioavailability of chemical species of other elements is, in general, difficult to predict.

Before bioavailability becomes relevant, substances must be accessible to the living organism at risk. An extreme case would be where a substance occurs in a part of the globe where life is impossible. In such circumstances, the substance is not accessible to any life form. The cases that require careful consideration are those where living organisms occur, but where essential nutrients or toxicants are locked into physical or chemical compartments that inhibit or prevent contact with the living organisms. In such cases, bioaccessibility becomes the limiting factor that determines the possibility of exposure and resultant adequate nutrition or potential toxicity. This is particularly true in relation to substances contained in soils, sediments and other particulate matter. For particles, bioaccessibility requires consideration of the physical nature of the particle, of the distribution of substances of concern and particularly of whether they occur on the surface of the particles. Wherever a substance occurs in a particle, the chemical speciation of both the substance of concern and of the matrix and surface of the particle influences both accessibility and availability of the substance (see below). Figure 9 illustrates these considerations in the context of a unicellular organism living in an aqueous medium. More complex diagrams may be developed from this for multicellular organisms and other environments. Environmental factors may restrict or facilitate accessibility, as may many aspects of the biology of the organism, from behaviour, through physiology to fundamental molecular properties of receptors.

Determinants of Bioaccessibility and Bioavailability. Metallic elements in non-ionic and uncombined form are mostly not bioaccessible or bioavailable. Mercury vapour is a notable exception. Mercury vaporizes under normal environmental conditions and dissolves in cell membranes because the vapour is lipid-soluble. In the case of mercury, the cationic forms, mercuric and mercurous ions, are much less able to pass into or through biological membranes. In contrast, mercuric chloride exists in seawater in an unionized lipid soluble form that can be absorbed readily by living organisms. Further, mercuric chloride can be converted by various organisms into methylmercury chloride, which is readily absorbed by living organisms because of its lipid solubility and which is sufficiently stable to bioaccumulate and to biomagnify in food webs.

In general, metallic elements that form hydrated ions readily upon solution in water are accessible to living cells through this medium. This reflects their solubility products. Elements that form insoluble precipitates in water may not be bioaccessible except to organisms that can solubilize them after phagocytosis

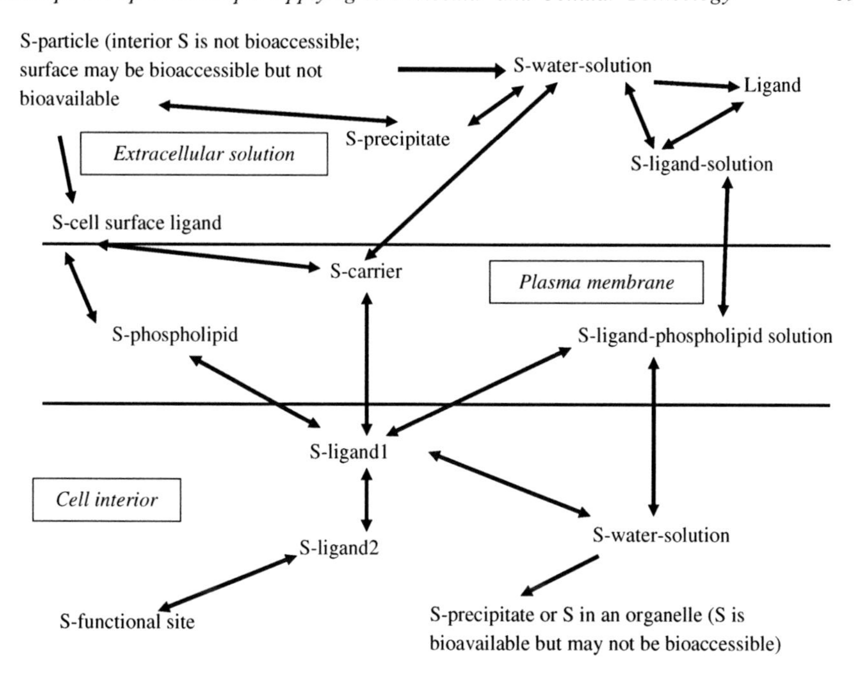

Figure 9 Bioaccessibility and bioavailability – schematic representation of the relationships between an extracellular substance, S, and the uptake of the substance by living cells.

or some similar activity. Apart from this, in general, the hydrolysis of metallic elements in water in the presence of dioxygen determines the bioavailability of the elements. For example, iron is found in natural aerobic waters in very small amounts because the redox potential restricts it to the Fe^{3+} state that precipitates as insoluble ferric hydroxide. Thus, iron cannot be found in natural aerobic waters as a soluble divalent cation, *e.g.* hydrated Fe^{2+}, or as a soluble anion such as FeO_4^{2-}. Hydrated ions may be bioaccessible but may not be bioavailable if there is no mechanism for their uptake. Absence of an uptake mechanism may reflect the size of the hydrated ion. Thus, a knowledge of relevant chemical speciation and the uptake mechanisms applicable to the organisms at risk is essential for any risk assessment of exposure to metallic elements. It should also be noted that metallic elements are not always bioavailable as hydrated cations. For example, chromium is bioavailable largely as the anion chromate. Unfortunately, the bioavailable form is often given in the scientific literature as Cr^{6+} and this has led to some confusion.

A substance may be complexed by inorganic and organic ligands, or adsorbed onto or bound within particles. The bioavailability of complexed ions varies with the nature of the complex. For example, aluminium complexed with citrate is more easily absorbed from solution than aluminium complexed with hydroxide, and is therefore more bioavailable. The absorption and bioavailability of many divalent ions is reduced by complexation with phytic acid but may be enhanced by complexation with some chelating agents such as ethylenediaminetetraacetic acid (EDTA). Often, aqueous complexes exist in an equilibrium state that may fluctuate considerably with environmental conditions. Such conditions may change quite rapidly, sometimes with dramatic consequences for the affected ecosystem.

In addition to being affected by complexation, metallic elements and ions derived from them can cycle between oxidation states of varying bioavailability, *e.g.*, as described earlier, ferrous ions are generally more bioavailable than ferric ions. Complexation and redox cycling are often associated with large differences in reactivity, kinetic lability, solubility and volatility because of resultant changes in chemical speciation. Thus, timing of exposure in relation to environmental conditions is an important factor to consider. In particular, chemical changes resulting from the contrasting effects of photosynthesis in the daytime (oxygen release, carbon dioxide fixation, increase in pH) and respiration at night (carbon dioxide release, oxygen uptake, decrease in pH) can cause major differences in water chemistry and hence in bioaccessibility and availability of both nutrients and toxicants.

In the simplest consideration of bioavailability, uptake into cells may be driven by diffusion or another form of electrochemical gradient. The concentration-of-substance or the charge that drive uptake through the plasma membrane will each be those near the membrane. Both may be dependent upon biotransformation and (or) localization within cells or cell compartments. Uptake is also affected by components of the exposure medium, such as the presence of similar chemical species that may compete for uptake sites. Bioavailability of an organic compound is often determined by its presence in solution in a surrounding aqueous medium and is dependent upon its hydrophobicity (lipophilicity). For ionizable organics, hydrophobicity is dependent on pH. For example, organic acids become unionized, and thus more hydrophobic, at acid pH. Notably, the dependence of total uptake simply on the aqueous concentration of the dissolved organic molecules of concern ceases to apply if the concentration of dissolved organics is kept constant by continuing dissolution from a large fraction sorbed to particulate matter. In such circumstances, uptake is continuous, perhaps at a slow rate, and large amounts of such organics may accumulate in exposed organisms with passage of time. Thus, time becomes the main determinant of total uptake.

Whatever the biologically available form of a substance in solution, it may be derived from other chemical species, and the thermodynamic equilibrium for production of an available solute or the rate of its release from a bound form may be the limiting factor for its uptake by living cells, *i.e.* transfer from the external medium to a biological receptor on or in the cell. Knowing which chemical species determine the rate and amount of uptake by living organisms of a substance of concern is essential for risk assessment. Thus, to determine the relevant chemical species for risk assessment of a substance, three questions must be answered:

1. What is the mechanism of uptake of each chemical species of the substance?
2. Which chemical species determine(s) the rate of uptake and excretion of the substance by the cell?
3. How do chemical species interact in the uptake and excretion processes? Interactions may occur outside the organism, where species may interact directly or compete for transport sites, or inside, where they may compete for binding sites that regulate transport systems.

Bioaccessibility and Bioavailability in Relation to Particulate Matter. Any substance must come in contact with a living organism before it can interact with it and be absorbed. Thus, a substance trapped inside an insoluble particle is not bioaccessible although substances on the surface of the same particle may be bioaccessible. Bioaccessible substances are not necessarily bioavailable. Substances on the surface of particles may be accessible to some organisms following phagocytosis but may not be accessible to organisms that require the substances to be in aqueous solution or which lack appropriate surface receptors. Thus, bioaccessibility, like bioavailability, is a function of both chemical speciation and biological properties.

In some cases, bioaccessibility will be the limiting factor that determines whether a substance is or is not toxic. This is particularly important in relation to substances in soils, sediments, aerosols and other particulate matter to which humans may be exposed.

Bioaccessibility and Bioavailability Through a Food Chain. Organisms can be placed in a chain of dependence, known as a food chain, with several different trophic levels (levels at which organisms feed). Such a chain (Figure 10) starts with plants or other primary producers absorbing light, carbon and other nutrients, and passing nutrients and organic molecules with their inherent chemical energy to higher trophic levels of herbivores and carnivores. Each trophic level produces waste material as excretory products and dead matter, and carbon dioxide from respiration. Decomposer organisms such as bacteria and fungi break down the waste products, releasing nutrients back to the environment where they are available for re-use. Thus, nutrients cycle between organisms and the environment. This is part of the general biogeochemical cycle that applies to all the elements used by living organisms. Integrity of biogeochemical cycles must be maintained to ensure continuing biological productivity.

Food chain bioaccessibility and bioavailability varies between elements, depending upon the environmental chemical species, the chemical species produced by the organisms involved, how they are stored, and the sinks for different chemical species in the environment. Thus, mercuric sulfate, which is

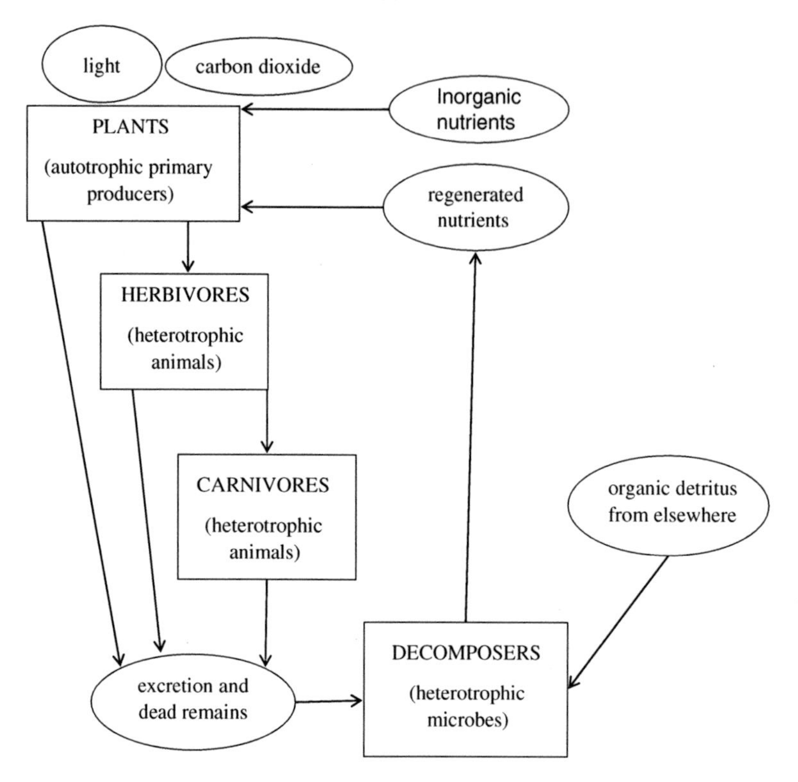

Figure 10 A simple food chain.

water-insoluble, tends to stay wherever it enters the environment. In contrast, methylmercury chloride, which is chemically stable and lipid soluble, is readily taken up by living organisms and passes from prey to predator until the ultimate predators in a food chain accumulate it to high levels, a process known as biomagnification. It is usually assumed that proportions of chemical species remain roughly constant in different compartments but this may not be true, especially in the aquatic environment. The aquatic environment is affected by changes from photosynthesis to respiration from day to night, with resultant changes in carbonates and pH. It is also affected by intermittent pollution incidents, or, in estuaries, by tidal flow, with resultant fluctuations in salinity and sediment disturbance.

Some organisms accumulate certain elements and compounds from the environment (bioconcentration, *e.g.* metals in plant tissues), and from their prey (bioaccumulation, *e.g.* dioxins in predatory organisms), causing them to have very high body loads relative to outside concentrations. Because of losses of organic matter owing to respiration, each successive trophic level in a predator–prey relationship usually has a lower biomass, defined as the mass of living material in a given area at one time, or a lower productivity than the levels below it. The body concentration of bioaccumulated substances passed up the food chain can therefore increase through the process of biomagnification, sometimes resulting in toxic doses to organisms higher up the chain.

All ecosystems have two types of food chain – the grazing food chain based directly on plant photosynthesis within the system, and the detritus food chain based on consumption of organic detritus by detritivores (organisms feeding on detritus) that are eaten by carnivores. Normally, the detritus chain is based on the waste products of the system's own resident organisms. An estuarine ecosystem differs from most others in its great reliance on imported waste from other systems.

Bioaccessibility and Bioavailability Through a Food Web. A food web is a more holistic concept than a food chain. Figure 11 gives a simplified soil food web. Even with such a simple web there can be a complex pattern of flow of energy, carbon, nutrients and toxicants based on the feeding preferences of different species, as indicated by the lines on the diagram. These feeding preferences determine bioaccessibility of toxicants that may be found in only a few species. In such cases, bioaccessibility depends on the feeding preferences of predators on these species. For any given habitat there is a degree of stability such that the same assemblages of species are present in the food web in successive years, with the same dominant and rare species. Similarly, the same flow pathways remain important while others are less significant.

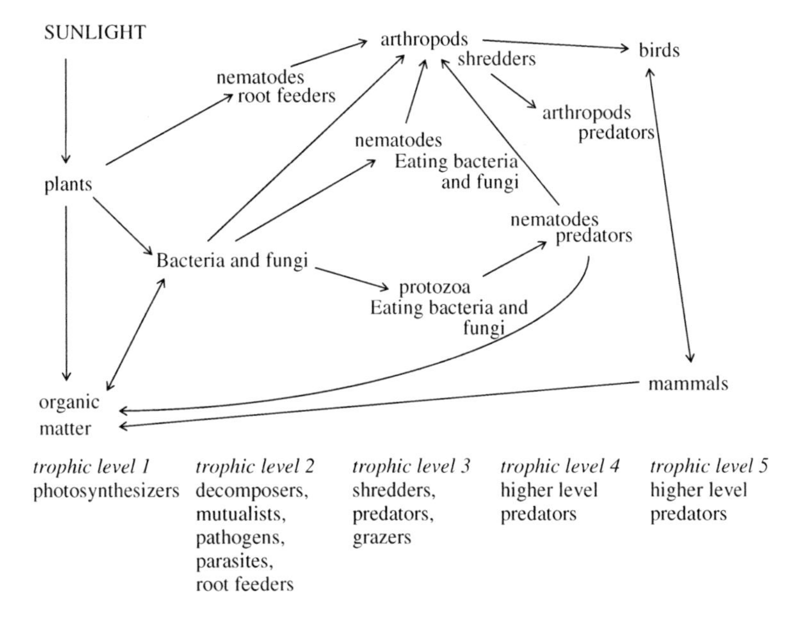

Figure 11 Soil food web showing relations between soil, organic matter, micro-organisms, plants, insects and so on, birds and mammals.

Bioavailability, Routes of Exposure and Absorption in Humans. Bioavailability to humans varies with routes of exposure. This is discussed in Section 3.2 (Absorption).

Further Reading

L. Shargel and A. B. Yu, *Applied Biopharmaceutics Pharmacokinetics*, McGraw-Hill, New York, 4th edn, 1999.

J. R. Dean, Bioavailability, *Bioaccessibility and Mobility of Environmental Contaminants*, Wiley Blackwell, Oxford, 2007.

R. Naidu, *Chemical Bioavailability in Terrestrial Environment (Developments in Soil Science)*, Elsevier Science, Amsterdam, 2008.

M. Nordberg, D. M. Templeton, O. Andersen and J. H. Duffus, Glossary of terms used in ecotoxicology (IUPAC Recommendations 2009), *Pure Appl. Chem.*, **81**, 829–970.

2.3 Carrier

carrier
1 Substance in appreciable amount which, when associated with a trace of a specified substance, will carry the trace with it through a chemical or physical process.
2 Gas, liquid, or solid substance (often in particulate form) used to absorb, adsorb, dilute, or suspend a substance to facilitate its transfer from one medium to another.

A carrier, in chemistry, refers to a substance that chaperones or carries another. In toxicology and toxicokinetics, it is used conventionally in several ways – subtly different, perhaps, but all related to the general concept. The parent definition mentions the carrier in 'appreciable' amount, and a referent substance in 'trace' amount. Without attempting a definition of these terms, we can note that a large difference in substance amount, perhaps a concentration gradient, is implied. The term is often used in describing preparations of radioisotopes, and this illustrates the concept nicely. If all the atoms of an element in a sample are radioactive, then the element is 'carrier-free'; otherwise, we would say that the cold isotope was a carrier. 'Carrying' in this sense refers only to dilution; a carrier here refers to a chemical species that changes the amount of another species from which it is distinguishable only by a particular analytical method. Combining the above two concepts, the occurrence of a large difference in amount of two substances that cannot be readily distinguished by available analytical techniques, we come to the concept of a carrier as a diluent.

In a different sense of the term, it is not dilution of one species by another but a physical interaction that is specified. The English verb 'carrying' is often used to describe a physical process of transport. Thus, if one molecule can interact with another, *e.g.* by adsorption or electrostatic attraction, one can be said to carry the other. It is in this sense that we refer to a carrier protein as interacting with a ligand in such a way that the ligand's distribution is determined by the protein. Again, concentration gradients would be important, but now from the point of mass action; a high concentration of the carrier will determine the relative distribution of the partner. This is what we mean, for instance, when we note that a carrier may compete with a substance's binding to a surface. A carrier substance (often a protein) is one that is used to deliver specifically another substance to an intended location, for instance, often referring to a substance that delivers a drug to its site of action.

In practice, we might add one substance to a sample to prevent another substance from being lost due to its presence in low quantities, *e.g.* by adsorption to a surface. The former substance would then be described as a carrier, and could contribute to the recovery of the latter by all the above processes – direct binding, competition for binding and dilution or displacement into a phase of interest. Albumin is often used as a 'carrier protein' because of its ready abundance and mixed hydrophobic and hydrophilic nature.

A final meaning of the term 'carrier' in biology is its use to describe a molecule or molecular complex that 'carries' a substance across a biological barrier. Here, the term 'transporter' is often substituted, though this may be a more specific term if it is intended to imply active transport requiring energy, as in the case of ATP hydrolysis, coupled transport or conformationally coupled processes. Such transport frequently refers to delivery across a cell membrane.

This brief discussion of physical meanings of carrier does not include 'carrier' as used in genetics.

See also Section 3.2 on absorption.

Further Reading

J. W. Kimball, Transport across cell membranes, in *Kimball's Biology Pages*, August 17, 2002 (cited January 2003). Available at <http://users.rcn.com/jkimball.ma.ultranet/BiologyPages/D/Diffusion.html>.

2.4 Biotransformation

biotransformation
Chemical conversion of a substance that is mediated by living organisms or enzyme preparations derived therefrom.

Chemical conversion refers to the transformation of one chemical species into another. The main point in using the term biotransformation is that the conversion must be carried out by (i) a living system or (ii) enzymes derived therefrom. In general, biotransformation refers to conversions that are carried out by enzymes. Conversion carried out by a living system includes all chemical conversions that go on within the cell or the body, either as part of the metabolism of endogenous substances or of exogenous xenobiotics, drugs, toxic substances, *etc*. However, interconversion of intermediates in a defined metabolic pathway would more usually be referred to as metabolism than as biotransformation. Thus, the term 'biotransformation' normally refers to a xenobiotic, unless the xenobiotic is a drug, in which case the term 'drug metabolism' would be preferred.

The biotransformation need not occur within the organism. For instance, microorganisms are often employed in waste management and environmental clean-up, *i.e.* in remediation. Including derivative enzyme preparations in the definition reinforces the intent that the term refers to enzyme-catalysed conversions. In principle, one could manufacture an enzyme by total chemical synthesis and then use it for chemical conversion of a substance; this would qualify as biotransformation. However, this proviso in the definition is really intended to include the more usual scenarios where a bacterial culture, a cell or tissue homogenate, or an enzyme purified therefrom, is used for a specific intended conversion. Biotransformations using purified enzymes or enzyme-enriched preparations are also useful in synthetic organic chemistry.

Biotransformation is one method of clearance of a substance, and it may result in a product of greater or lesser biological effect (toxic or therapeutic) than the starting material. In the therapeutic context, biotransformation can be exploited to facilitate delivery to a site of effect, and activation at that site. Dopamine, used in the treatment of Parkinson's disease, does not readily cross the blood–brain barrier. However, L-dopa (3-hydroxy-L-tyrosine) does, and is decarboxylated to dopamine in the central nervous system. The cell membrane is another important barrier. An acetoxymethyl ester (AM) derivative is often employed to mask a carboxylate functionality. Once inside the cell, endogenous esterase activity hydrolyses the ester, and the charged parent acid is trapped within the cell by virtue of its negative charge. Chemotherapeutic agents are also designed with regard to biotransformations that might take place in the unique environment of the neoplastic tissue, based, for example, on the hypoxic nature of its core.

Biotransformation is an extremely important concept in toxicology, as it often involves systems that have evolved to detoxify xenobiotics. However, initial steps in detoxification may increase toxicity. An example would be epoxidation of an aromatic hydrocarbon, increasing its solubility for excretion or creating functionality for conjugation, but in the process creating reactive chemical species that can damage biomolecules like lipids, proteins and DNA by creating oxygen-centred radicals and forming adducts.

The above processes are the basis for a classification of reactions of drug metabolism into phase I and phase II. Phase I refers to the chemical modification of a substance by processes such as hydroxylation, oxidation, reduction or chlorination. This introduces functionality for subsequent further metabolism.

Phase II refers to conjugation of the phase I products, usually enhancing their hydrophobicity and facilitate urinary or biliary excretion. Whereas phase I processes increase hydrophilicity and add functionality at the risk of increased toxic potential, those of phase II almost invariably decrease toxicity.

Phase I Metabolism. Phase I metabolism includes various reactions that modify the core structure of the xenobiotic. These include oxidations by mixed-function oxidases (MFOs), reductions, hydrolyses, hydrations and isomerisms. The MFO cytochrome P_{450} (P450) is so important for phase I oxidations that it is usual to consider oxidations by P450 separately from others. Non-P450 oxidations are catalysed by enzymes such as alcohol dehydrogenase, xanthine oxidase, amine oxidases and aromatases; they require oxygen. Reductive biotransformation includes reactions that are generally inhibited by oxygen and require NADPH. Substrates include azo- and nitro-compounds, halogenated compounds, nitrogen heterocycles and epoxides. Several enzymes effect hydrolysis, most significantly the esterases, while hydration without hydrolytic cleavage is the domain of less abundant hydratases. Simple reactions such as cyclization, isomerization, dimerization, transamidation and decarboxylation are also considered phase I.

The P450 family of enzymes uses NADPH and molecular oxygen to hydroxylate many different substrates. These are heme-containing enzymes with molecular weights (relative molar masses) of about 45–55 kDa embedded in the membrane of the endoplasmic reticulum. The P450 genes are classified into families and subfamilies, and well over 30 human P450 proteins are uniquely identified. Some of these are primarily important for drug metabolism (*e.g.* P450s 2E1, 2D6, 2C9, 3A4); others function in sterol synthesis (*e.g.* P450s 11A1, 21A2). Genetic variation in P450 isoforms is associated with differing rates of drug metabolism (*e.g.* P450 2D6 variants give rise to different rates of debrisoquine metabolism), and potential susceptibility to some carcinogens.

Phase II Metabolism. Phase II metabolism acts upon the products of phase I to render them even less toxic and more water-soluble by conjugation reactions, enhancing urinary and biliary excretion. These reactions include addition of glucuronic acid (glucuronidation), glycosylation, sulfation, methylation and acetylation. Conjugation of many substances (generally strong electrophiles produced by phase I metabolism) to glutathione represents a major route of detoxification. This may occur spontaneously, or be catalysed by glutathione *S*-transferases. The glutathione (Gly-Cys-Glu) thioconjugate may be excreted directly through the bile or urine, or may become a substrate for sequential attack by glutamyl transpeptidase and a second peptidase that results in release of the cysteine conjugate of the xenobiotic. This occurs particularly in the liver and kidney. Conjugations with fatty acids or cholesterol esters also occur, as with the conjugation of cannabinoids to stearic and palmitic acids. Glucuronidation of bilirubin is important in preventing build-up of this potentially toxic metabolite. In general, conjugations require an 'activated' intermediate, either of the xenobiotic itself or involving UDP-glucuronic acid (for glucuronidation), *S*-adenosylmethionine (for methylation), acetyl coenzyme A (for acetylation) or phospho-adenosine phosphosulfate (for sulfation). Though not generally commonly used, the term 'phase III metabolism' refers to metabolism of conjugates arising from products of phase II. An example would be the further metabolism of glutathione conjugates in the gut.

Among the most important sites of biotransformation from the perspective of toxicokinetics are the liver and the lung, although the gastrointestinal tract, kidney and erythrocytes are also important in xenobiotic metabolism. The liver plays a central role in biotransformation and detoxification of xenobiotics; indeed, that is one of its major roles. The 'first pass effect' refers to the fact that oral exposure to a substance absorbed in the gut will first meet the liver through the portal circulation, and undergo phase I processing. Most of the enzymes of phase I and phase II metabolism were initially identified in liver tissue. Phase I oxidative enzymes in liver and other cells are almost exclusively in the endoplasmic reticulum, whereas phase II enzymes, including the glutathione *S*-transferases, are cytosolic: an exception is glucuronosyl transferase, which is in the endoplasmic reticulum. There are effective transfer systems for sulfate, glucuronic acid, glutathione and glycine conjugates in liver cell membranes. These permit uptake of conjugates into liver cells and their secretion into bile. Similar transfer systems are found in kidney cell membranes.

The lung is the portal of entry for inhaled substances and has its own defensive barrier. Pulmonary alveolar macrophages are adapted to antioxidant defence. Paradoxically, the very defences of lung tissue result in metabolic activation of many compounds. This is the consequence of a metabolically active, aerobic tissue. Many toxicants first enter the body into the circulation. The erythrocyte, as the major circulating cell, protects itself by detoxification through biotransformation. Erythrocytes lack nuclei and reproductive capacity, but nevertheless serve a unique role in detoxification. They exist in a high O_2 and high iron environment, and must withstand this environment of high oxidative stress. Iron, haemoglobin and oxygen (oxidative stress) participate in oxidative biotransformation reactions in the erythrocyte. Glutathione conjugates are actively transported outward across the red-cell membrane. A major protective enzyme against oxidative stress in erythrocytes is glutathione peroxidase. Characteristic biotransformations in erythrocytes are glutathione transferase reactions and N-oxidations of some substrates.

Further Reading

A. Parkinson and B. W. Ogilvy, Biotransformation of xenobiotics, in *Casarett & Doull's Toxicology – The Basic Science of Poisons*, ed. C. D. Klaasen, McGraw-Hill, New York, 7th edn, 2008, p. 161.

National Library of Medicine (Content source): Emily Monosson (topic editor), 'Biotransformation,' in *Encyclopedia of Earth*, ed. C. J. Cleveland, Environmental Information Coalition, National Council for

Science and the Environment, Washington, D.C. (First published in the *Encyclopedia of Earth* April 13, 2007; last revised February 28, 2008; retrieved August 21, 2008). Available at <http://www.eoearth.org/article/Biotransformation>.

2.5 Reactive Oxygen Species

reactive oxygen species (ROS)
Intermediates in the reduction of molecular O_2 to water. *Examples*: superoxide ($O_2^{-\bullet}$), hydrogen peroxide (H_2O_2) and hydroxyl ($HO^\bullet$).

Cells produce energy from mitochondrial respiration, which involves a stepwise, four-electron reduction of molecular oxygen. In defining reactive oxygen species (ROS) as the intermediates that occur in the biological reduction of molecular O_2 to water, we acknowledge the central role of aerobic respiration in human biology. At the same time, we restrict the term to several oxygen species, each with its own distinct chemistry and important role in biology. Because no chemical process is completely efficient, the intermediates in respiration necessarily 'leak' from the reaction pathway to some extent, and serve as the cell's major source of exposure to ROS.

Reduction of Dioxygen. Figure 12 depicts the stepwise reduction of O_2. Molecular oxygen, or dioxygen (O_2), exists in a double-bonded triplet ground state with an excess of two π-bonding electrons. Acceptance of the first electron into an antibonding orbital reduces the bond strength and creates the radical anion species superoxide ($O_2^{-\bullet}$). A second electron, also accepted into a π antibonding orbital, further reduces the bond strength, and produces the singly bonded peroxide species (O_2^{2-}). Because the pK_a values of this species are >14 and 11.8, it exists as H_2O_2 under biological conditions. The next electron, also accepted into an antibonding orbital, breaks the single bond, and the resulting $HO^\bullet$ is reduced to hydroxide and water. The one-electron redox potentials shown in Figure 12 are useful for determining whether a step will occur spontaneously in the presence of a redox-active metal complex. This principle has been used, for example, in designing chelating agents for iron that will avoid catalytic generation of $HO^\bullet$.

The traditional descriptor 'reactive' is somewhat misleading; for a radical, superoxide, is relatively stable and persists to diffuse into the extracellular space. For instance, reperfusion injury (see below) can be diminished by scavenging superoxide from the extracellular space. The $HO^\bullet$ radical, on the other hand, reacts with carbon–carbon bonds at diffusion-controlled rates, and so is an extremely harmful species. Peroxide is mainly harmful due to

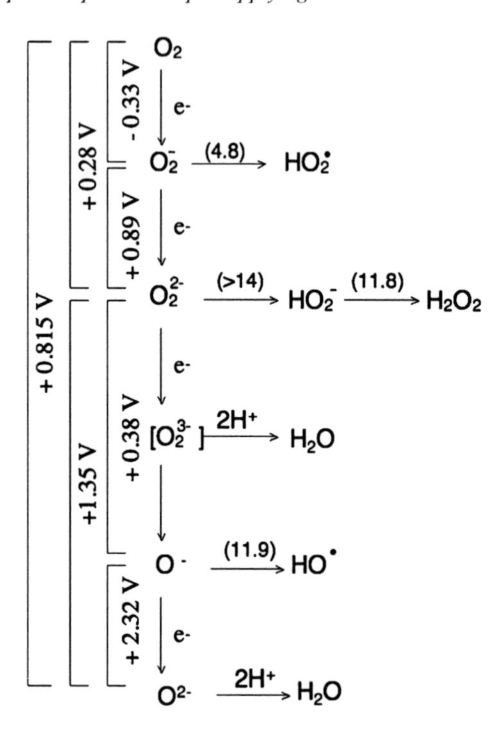

Figure 12 Four-electron stepwise reduction of molecular oxygen to water. Numbers in parentheses are pK_a values for protonation of the intermediates. Redox potentials for various reductions are shown on the left-hand side in volts at pH 7.

generation of $HO^{\bullet}$ either through homolytic ($HOOH \rightarrow 2HO^{\bullet}$) or reductive heterolytic ($HOOH + e^{-} \rightarrow HO^{\bullet} + HO^{-}$) cleavage of its O–O bond.

The reduction of molecular oxygen to water occurs not only in the mitochondrion where it is coupled to oxidative phosphorylation, but also by the action of oxidative enzymes in the endoplasmic reticulum, lysosomes, peroxisomes and even in the cytosol. While leakage of intermediates from respiration is a major source of intracellular ROS in aerobic organisms, other sources are also significant. Absorption of high-energy electromagnetic radiation (*e.g.* X-ray or UV) can permit radiolysis of water to yield $HO^{\bullet} + H^{\bullet}$. Bursts of superoxide occur when neutrophils are activated by appropriate stimuli that occur during inflammation. This involves an NADPH oxidase activity that uses a protein complex known as cytochrome b_{558} to shuttle electrons from within the cell to reduce O_2 to superoxide at the cell surface. This machinery also exists in some other cells. In addition, during purine metabolism, the enzyme xanthine oxidase uses O_2 as an electron acceptor in the conversion of hypoxanthine into uric acid, thus generating superoxide.

Transition metals with redox potentials in a biologically accessible range, such as iron and copper, can accept and donate electrons in a catalytic fashion.

$$1)\ O_2^{-\bullet} + Fe^{3+} \longrightarrow O_2 + Fe^{2+}$$

$$2)\ Fe^{2+} + H_2O_2 + H^+ \longrightarrow Fe^{3+} + H_2O + HO^\bullet$$

$$O_2^{-\bullet} + H_2O_2 + H^+ \longrightarrow O_2 + H_2O + HO^\bullet$$

Figure 13 Iron-catalysed Fenton reaction. The overall reaction (below the line) is the sum of reactions 1 and 2.

The Fenton reaction with ROS (Figure 13) is extremely important because it generates $HO^\bullet$; it is a catalytic cycle, and available iron can generate large amounts of reactive $HO^\bullet$ quickly.

Adverse Effects of ROS. ROS are important in toxicology because of the cellular and molecular structures they target. On the structural level, major targets are the cell membrane (determining cellular integrity), mitochondria (not only providing the cell with energy, but determining its fate through the role the mitochondrial permeability transition pore and cytochrome *c* release play in apoptosis) and the nucleus. On a molecular level, this is characterized by damage targeted to lipids, proteins and nucleic acids. Double bonds in lipids of the bilayer membranes of cells and organelles are subject to attack by ROS (especially $HO^\bullet$). The lipids then form peroxides that themselves propagate the injury. A consequence of losing internal membrane integrity is the inability to control ion and water fluxes. Oxidative damage to proteins ranges from indirect effects, such as the formation of unnatural disulfide bonds in an oxidizing atmosphere, to direct poisoning of enzymes through attack of essential residues. Oxidation of many proteins also targets them for proteosomal degradation. Nucleic acid mutations can lead to short- and long-term effects. Short-term effects include changes in gene expression and cell phenotype that may be overcome with rapid activation of DNA repair mechanisms. Longer-term effects include either apoptosis or the potential for malignant transformation if genetic defects are not repaired.

ROS are rarely discussed without reference to the many defence mechanisms that have evolved to protect cellular structures against them. These include: (i) enzymes such as superoxide dismutase and catalase that eliminate superoxide and peroxide, respectively; (ii) antioxidants such as ascorbate and vitamins E and A; (iii) the glutathione/glutathione peroxidase system; and (iv) proteins that sequester potentially Fenton-active metals, such as ferritin for iron or metallothionein for copper. Despite these defences, excess production of ROS is harmful. It is a major contributor to long-term tissue damage in diseases such as haemochromatosis, where excess accumulation of iron results in increased Fenton activity. Another process in which overproduction of ROS causes extensive tissue damage is reperfusion injury. Following infarction of a tissue, restoration of perfusion with oxygenated blood results in a rapid production of ROS, notably superoxide, and tissue damage is exacerbated.

Damage following a myocardial infarction, for example, can be decreased with radical scavengers, antioxidants, metal chelators, superoxide dismutase and catalase.

Though generally thought of as harmful in the field of toxicology, ROS are an essential part of normal cell function; they serve as rapid, diffusible signalling molecules. For example, basal levels of H_2O_2 are required for signalling by the platelet-derived growth factor. Generation of superoxide at the surface of activated neutrophils is necessary for efficient killing of invading bacteria. Patients with genetic defects in the superoxide-generating cytochrome b_{558} machinery develop constrictive granulomas, especially in the lungs (chronic granulomatous disease) as a result of repeated and persistent infections.

The term 'oxygen-centred radical' should be avoided. Simple molecular radical species containing nitrogen are also of biological importance, and include the NO^-, $NO^•$ and NO^+ species. Without distinction of where electron density is centred, these are sometimes referred to as reactive nitrogen species. They are important in regulating vascular physiology by eliciting smooth muscle relaxation, and their metabolism includes protein thionitrosylation and formation of peroxynitrite.

Further Reading

R. B. Mikkelsen and P. Wardman, Biological chemistry of reactive oxygen and nitrogen and radiation-induced signal transduction mechanisms, *Oncogene*, 2003, **22**, 5734.

2.6 Aerobic and Anaerobic

aerobic
Requiring dioxygen.
anaerobic
Not requiring dioxygen.
Antonym: aerobic.

The term 'aerobic' refers to the requirement of an organism for dioxygen (O_2), and therefore has a basis in metabolism. A description of our current knowledge of metabolism can be found in any good biochemistry textbook. One that is also accessible through the internet is Boyer's *Concepts in Biochemistry* (*see* Further Reading). Central to understanding aerobic metabolism is understanding the mechanism that higher organisms use to generate adenosine triphosphate (ATP) in the process of respiration. Respiration, in turn, refers to the oxidation of fuel molecules with dioxygen (with the exceptions noted below) as the ultimate electron acceptor and the derived energy stored in the terminal phosphate bond of ATP (oxidative phosphorylation). The breakdown of fuel molecules to yield metabolic energy in the absence of dioxygen is referred to as anaerobic metabolism, and is considered to be the antonym of aerobic metabolism. Here, a fuel molecule (very frequently glucose) is split into two moieties in a process called fermentation. One moiety then oxidizes the other and the energy released by the reaction is used for generating ATP. In higher organisms, this is typified by glycolysis, which produces ATP under conditions of relative dioxygen limitation, but also serves as an entry point to oxidative metabolism by providing pyruvate to the tricarboxylic acid (TCA) cycle.

Many cells can perform aerobic metabolism when dioxygen is present but survive by anaerobic metabolism when it is not. These are called facultative anaerobic cells. Cells that survive only by anaerobic metabolism, *i.e.* cells that do not have the capacity to utilize dioxygen as an oxidant and for whom dioxygen is therefore toxic, are called obligate anaerobes.

Biochemistry of Aerobic Metabolism. In the mitochondria of cells of higher animals, including humans, a cyclic series of enzyme reactions is central to the production of reducing equivalents for respiration. This is known variously as the TCA cycle, Krebs cycle or citric acid cycle (Figure 14).

Various nutrients (carbohydrates, amino acids and fatty acids) are metabolized by distinct pathways to generate two-carbon acetate units that are coupled to coenzyme A (CoA). The resultant acetyl coenzyme A reacts with the four-carbon substrate oxaloacetate to produce the six-carbon citrate. Sequential

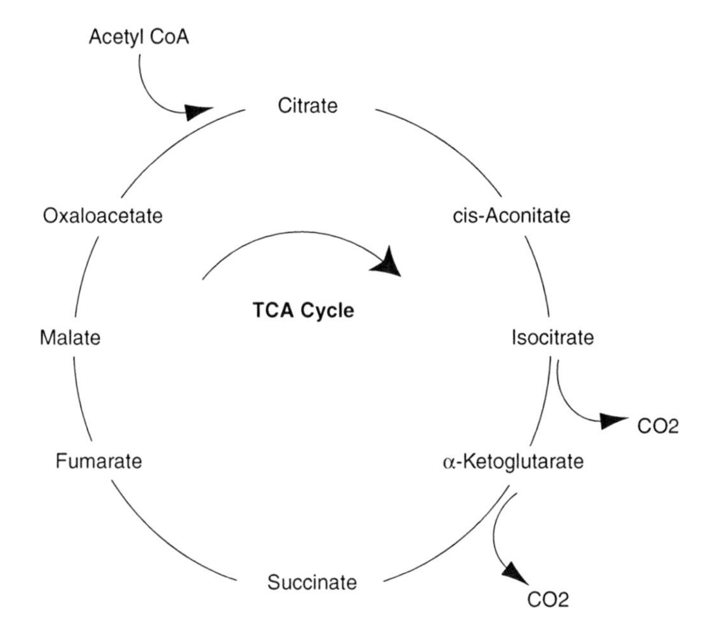

Figure 14 Tricarboxylic acid (TCA) cycle.

enzymatic conversion of citrate into isocitrate *via cis*-aconitate provides the substrate for isocitrate dehydrogenase, which converts isocitrate into 2-keto-glutarate with loss of CO_2. Loss of a second molecule of CO_2 converts 2-ketoglutarate into the four-carbon succinate. Oxaloacetate is regenerated from succinate *via* the intermediates fumarate and malate, to begin the cycle again. One turn of the cycle releases eight H atoms. These provide four pairs of electrons, which are recovered by the reduction of three molecules of nicotinamide adenine dinucleotide (NAD) to three molecules of $NADH_2$, and reduction of one molecule of flavin adenine dinucleotide (FAD) to $FADH_2$. If the source of acetyl coenzyme A is carbohydrate, two additional H atoms are released during conversion of pyruvate into acetate with production of an additional $NADH_2$. The electrons from $NADH_2$ and $FADH_2$ are passed down a reduction potential gradient known as the electron transport chain, consisting of a series of quinone- and cytochrome-containing proteins. These serve to split the electron pairs derived from $NADH_2$ or $FADH_2$ using the quinone/semiquinone and Fe(III)/Fe(II) redox couples, and achieve the stepwise four-electron reduction of O_2 to $2H_2O$. Reduction of one molecule of dioxygen to $2H_2O$ by $2NADH_2$ is coupled in the electron transport chain to the production of six molecules of ATP from ADP. This process is called oxidative phosphorylation. ATP production is coupled to electron transfer at three sites in the electron transport chain where the drop in potential is large enough to pump a pair of protons outward across the mitochondrial membrane, creating an electrochemical gradient or protonmotive force. Their return through a proton channel in an ATPase provides the energy to reverse the ATPase activity and act as ATP synthase. Toxic substances that allow the electron transport chain to continue with the reduction of dioxygen, but prevent the formation of ATP and dissipate the energy as heat instead, are called uncoupling agents. Examples include 2,4-dinitrophenol (DNP), carbonyl cyanide *p*-[trifluoromethoxy]phenyl hydrazone (FCCP) and carbonyl cyanide *m*-chlorophenyl hydrazone (CCCP).

Operation of the electron transport chain is one of the major sources of production of reactive oxygen species (ROS; see Section 2.7) in respiring cells. Increased flux of glucose through glycolysis and ultimately of the resulting pyruvate through the TCA cycle feeds an abundance of $NADH_2$ into the electron transport chain, and experimental hyperglycaemia leads to an increase in ROS that is thought to account for many of the chronic changes in diabetes. Overexpression of the uncoupling agent uncoupling protein-1 (UCP-1) dissipates the trans-mitochondrial membrane proton gradient required for operation of ATP synthase, and decreases many of the biochemical changes associated with the microvascular complications of diabetes.

Glycolysis and Fermentation. As noted above, cells that survive in the absence of dioxygen break down a nutrient molecule into fragments in a process called fermentation. One fragment then serves as the electron acceptor in the oxidation of another, and this oxidation is coupled to the formation of ATP. The most important nutrient for anaerobic metabolism is glucose.

Because fermentation results in incomplete breakdown whereas respiration degrades glucose to CO_2 and H_2O, the amount of energy released in respiration is much greater. Early life evolved under anaerobic conditions, but when O_2 became available in the environment, some organisms adapted to its use. Because of the greater efficiency of respiration, organisms today that can use either respiration or fermentation generally prefer respiration when dioxygen is available. Yeast and bacteria can be classified according to their fermentation products of glucose. For example, some break down glucose to ethanol and CO_2, others to ethanol and acetic acid, yet others to acetone or butanol. The fermentation of glucose to pyruvate is called glycolysis (Figure 15). This is the form of anaerobic metabolism that occurs in higher animals.

Glycolysis follows a regulated enzymatic pathway simpler than that of respiration. Overall, six-carbon glucose is split into two three-carbon molecules of lactic acid, which does not represent a redox reaction. However, before cleavage, glucose undergoes sequential phosphorylation and isomerization reactions and then is converted by aldolase into the two three-carbon moieties dihydroxyacetone phosphate and glyceraldehyde-3-phosphate. Glyceraldehyde-3-phosphate undergoes a multistep conversion into pyruvate, which then acts as the electron acceptor for oxidation of additional glyceraldehyde-3-phosphate, itself being reduced to lactate by lactate dehydrogenase. The oxidation of glyceraldehyde-3-phosphate by pyruvate is coupled to the generation of two molecules of ATP from ADP. Because dihydroxyacetone phosphate and glyceraldehyde-3-phosphate are interconvertible by the enzyme triose phosphate isomerase, both three-carbon fragments of glucose are ultimately converted into pyruvate. Two ATP molecules are required for the initial hexose phosphorylation reactions, and the subsequent three-carbon moieties yield two ATP each, for a net production of two ATP per glucose. The fate of pyruvate depends on the metabolic context of the cell (Figure 15). In anaerobic glycolysis it is converted into lactate. In alcoholic fermentation it is converted into ethanol with release of CO_2. Under aerobic conditions, it is decarboxylated and coupled to coenzyme A (CoA) with the release of CO_2 and generation of NADH from NAD^+. The resultant acetyl CoA enters the TCA cycle *via* reaction with oxaloacetate, while the NADH ultimately transfers its electrons to O_2 through the mitochondrial electron transport chain.

Physiological Considerations. In an aerobic cell, lactate is further oxidized to CO_2 and H_2O, and pyruvate is shuttled into the TCA cycle. Human cells can survive varying periods of anoxia, depending on their glucose or glycogen stores, and their complement of glycolytic enzymes. The availability of dioxygen can also affect the pattern of drug metabolism. Whereas phase I metabolism in the mammalian liver and other drug-metabolizing organs is typified by oxidation carried out by enzymes such as the cytochrome P450 and flavin-containing monooxygenases, monoamine oxidase and peroxidases, reductive phase I metabolism in other situations involves enzymes such as reduced cytochrome P450 and NADPH-cytochrome P450 reductase.

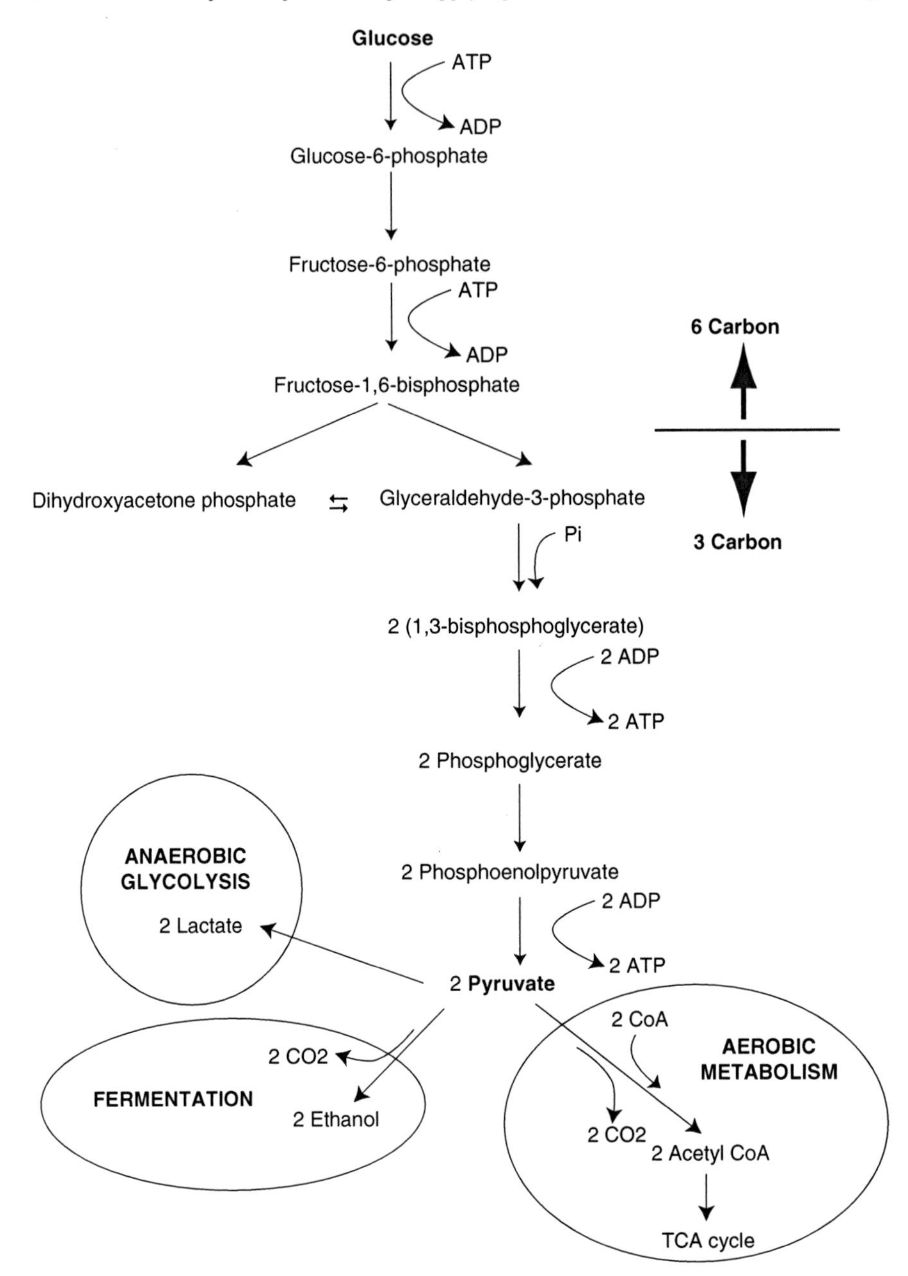

Figure 15 Glycolysis – the conversion of glucose into pyruvate with the net generation of two ATP. The point where a six-carbon structure is cleaved into two three-carbon structures is indicated.

Skeletal muscle has a high capacity for anaerobic metabolism, allowing prolonged periods of exercise. During exertion, the quantity of glucose metabolized in glycolysis may exceed the capacity of the oxygen supply to support mitochondrial reoxidation of NADH. Under these circumstances, a debt of lactic acid is built up from conversion of pyruvate. This lactic acid must be metabolized during recovery. Some is oxidized in muscle, while some diffuses out of the muscle cell and is converted into glucose in the liver or oxidized in heart and other tissues. In contrast to skeletal muscle, cardiac muscle is abundant in mitochondria and requires a constant supply of dioxygen. It has limited capacity for anaerobic metabolism, oxidizing its own pyruvate as well as oxidizing lactate released from other tissues. The enzyme lactate dehydrogenase is a critical determinant of this difference between cardiac and skeletal muscle, which catalyses the interconversion of lactate and pyruvate. The tetrameric enzyme is made up of any combination of the isoenzyme subunits H (abundant in heart) and M (abundant in muscle). The H subunit operates much more effectively with lactate as a substrate, compared to pyruvate, and is inhibited by high concentrations of pyruvate. Thus, in tissue rich in H subunits, such as heart, pyruvate is preferentially utilized by reactions other than that catalysed by lactate dehydrogenase, and the oxidation of lactate to pyruvate is facilitated. The M subunit is not inhibited by pyruvate, and converts pyruvate into lactate when means of oxidizing NADH and pyruvate are compromised.

Microbiology. The grouping of bacteria according to the effects of dioxygen on their metabolism and growth is important for their isolation and identification, as well as for understanding pathogenesis. One classification distinguishes among:

1. Obligate aerobes that require dioxygen and lack capacity for significant fermentation. Examples include the tubercle bacillus and some spore-forming bacilli.
2. Obligate anaerobes that grow only in the absence of dioxygen. While dioxygen inhibits some of the enzymes required by these organisms for fermentation it is also lethal to many. This is largely due to the lethal effect of superoxide in organisms lacking superoxide dismutase. Examples include *Clostridia* and *Propionobacter*.
3. Aerotolerant anaerobes, which grow in the presence or absence of dioxygen, but retain fermentative metabolism in both circumstances. These include most lactic acid bacteria.
4. Facultative organisms, which grow in the presence or absence of dioxygen, usually preferring the more efficient aerobic metabolism when dioxygen is present. These include enterobacteria and many yeasts.

While animal physiologists think of respiration as requiring dioxygen, microbiologists distinguish fermentation from respiration by determining whether the ultimate electron acceptor is organic (fermentation) or inorganic (respiration). Although most respiring bacteria use dioxygen as the acceptor, anaerobic respirers exist and include those that use nitrate (denitrifiers), sulfate (*Desulfovibrio*) or carbon dioxide (methane bacteria and some *Clostridium* sp.) as alternative oxidizing agents.

Anaerobic infections are caused by anaerobic bacteria, and occur in low dioxygen environments such as deep wounds and internal organs. They are frequently of mixed bacterial species and are characterized by abscess formation, foul smelling exudates and necrotic tissue destruction. Anaerobes normally colonize bodily regions of low dioxygen exposure, including the mouth (dental infections are commonly anaerobic), the gastrointestinal tract and the vagina (a danger in septic abortion). They cause infection when normal tissue barriers are damaged by injury, including surgical procedures. They are also found in soil and decaying vegetation. Common diseases caused by anaerobic *Clostridium* bacteria are gas gangrene (*C. perfringens*), tetanus (*C. tetani*) and botulism (*C. botulinum*). Because their growth is limited by dioxygen, infectious colonies are generally slow-growing and therefore difficult to treat.

Further Reading

R. F. Boyer, *Concepts in Biochemistry*, John Wiley & Sons, Inc., Hoboken, 3rd edn, 2006. Available at <http://bcs.wiley.com/he-bcs/Books?action=index&itemId=0471661791&itemTypeId=BKS&bcsId=2852>.

2.7 Mutagenicity

mutagenicity
Ability of a physical, chemical, or biological agent to induce (or generate) heritable changes (mutations) in the genotype in a cell as a consequence of alterations or loss of genes or chromosomes (or parts thereof).

Our heritable information consists of deoxyribonucleic acid (DNA) base sequences, or genes, that code for proteins and are arranged end-to-end with much intervening non-coding DNA. The linear arrangement of genes and non-coding DNA constitutes a chromosome, and in humans the genetic information of the nucleus is packaged into 23 pairs of chromosomes plus the X and Y sex chromosomes. In addition, mitochondrial DNA sequences encode information that is processed within the mitochondria, and may also be subject to mutation. The nuclear chromosomes are coated with various proteins, including histones and an additional amount of non-histone proteins that help package them into complex tertiary structures. This is discussed further above in Section 2.5. When an alteration in the readable sequence of a gene occurs, changes in the transcript of the gene will result and an abnormal, generally malfunctioning protein will be produced, or the protein may even be absent. Such changes, called mutations, drive evolution, but beneficial mutations are negligible in number compared to mutations that harm the organism. In addition to changes in sequence, whole segments of genes or even sections of chromosomes may be deleted or rearranged, and these changes are also referred to as mutations, although the term clastogenic is used to refer to the subset of mutations that result from chromosomal breaks and rearrangements. In contrast, changes in gene expression that do not result from changes in sequence are called epigenetic. Mutagens that lead to cancer are called carcinogens (Section 2.8); those that cause birth defects or malformations in the offspring are referred to as teratogens (Sections 3.11 and 3.12).

Mutations can arise spontaneously from errors in DNA replication or DNA repair. However, many are caused by agents that increase the frequency of mutations by interacting with, or chemically modifying, DNA, and these agents are referred to as mutagens. Many mutagens are chemical substances, but ionizing and ultraviolet radiation can also alter DNA structure physically (*e.g.* by inducing thymine dimer formation that interferes with transcription), and is also considered a mutagen. Some chemical mutagens intercalate with DNA, thus interfering with the fidelity of transcription. These are generally planar hydrophobic structures such as acridine orange or ethidium bromide that fit in the major or minor groove of the DNA helix (intercalation). The anticancer drug cisplatin is an inorganic compound that intercalates with DNA. Alkylating agents, on the other hand, form covalent adducts with DNA bases that result in faulty transcription. Commonly, a site of attack of monoalkylation is the 7-N position of guanine. Dialkylating agents can cause DNA strand crosslinking through this site. Alkylating agents include many anti-cancer agents such as nitrogen mustards (cyclophosphamide, chlorambucil), alkyl sulfonates (busulfan) and nitrosoureas. Base substitution in RNA or DNA with analogues such as halogenated pyrimidines, and base deamination with nitrous acid, are further examples of actions of mutagens. Some substances require activation within the cell, *e.g.* by cytochrome P450, before they become mutagenic. An example here is the activation of benzopyrene to its epoxide by CYP1A1.

Several types of mutations are well documented. A change in one base pair may result in a new triplet codon that codes for a different amino acid in the final protein, giving rise to a dysfunctional protein product. This is a missense mutation. A nonsense mutation occurs when the single base pair change results in a stop codon and a truncated product. One or more base pairs may be inserted into (insertion mutation) or deleted from (deletion mutation) the DNA sequence, resulting again in a different or absent protein product. If the insertion or deletion is not a multiple of three base pairs, the reading frame of codons will be changed and this is called a frameshift mutation. Duplication occurs when a DNA sequence is erroneously copied more than once during chromosomal replication. Duplications of genes such as *myc* and *Flt3* are often associated with a poor prognosis in cancer patients. If multiple copies of a gene result, this may result in the overexpression of the encoded protein. Expansion refers to a sequence of DNA that is replicated multiple times within a gene. For example, the sequence GAA in the first intron of the *FRDA* gene encoding frataxin has been found to be repeated up to 1700 times in patients with Friedreich's ataxia.

Heritability refers to the passing on of a DNA sequence from a cell to its daughters, and, if they give rise to gametes, to the progeny of these gametes. Thus, a mutation that becomes embedded in the germ line is passed on from one generation to the next, in either an autosomal or X-linked manner in humans, depending on whether the mutation is carried on one of the sex chromosomes. If a mutagen causes an alteration in DNA sequence of a non-germ line cell, division of that cell in an organism may result in dysfunctional or neoplastic tissue. It is now well recognized that mutations in mitochondrial genes may also be inherited.

Testing for the mutagenic potential of a substance is often carried out with the Ames test. In this test, several strains of *Salmonella typhimurium* are used. These strains have been rendered deficient in histidine synthesis through both deletion and frameshift mutations in the genes involved in histidine biosynthesis. After exposure to a substance, the occurrence of *S. typhimurium* colonies growing on histidine-deficient medium is an indication of the ability of the substance to have produced reversing or counteracting mutations of one or more of the different types. Rat liver microsomes may be included to allow for the possible requirement of bioactivation before the substance becomes mutagenic. This test is very sensitive, but it is also recognized as not necessarily predictive for eukaryotes. Further, the sensitivity of the Ames test is improved by using *Salmonella* strains deficient in DNA repair mechanisms, which at the same time potentially exaggerates the mutagenicity of test substances. Tests with whole animals, such as the Legator test, have been mentioned briefly in Section 1.1. They may be more realistic, in that host factors may also modify the mutagenic potential of a substance, but end points are less well-defined and the high-throughput feature of a screening test is lost.

Further Reading

B. B. Gollapudi and G. Krishna, Practical aspects of mutagenicity testing strategy: an industrial perspective, *Mutat. Res.*, 2000, **455**, 21.

IPCS, *Transgenic Animal Mutagenicity Assays*, EHC 233, WHO, Geneva, 2006. Available from International Programme on Chemical Safety (IPCS) INCHEM (Chemical Safety Information from Intergovernmental Organizations), 2008, at <http://www.inchem.org/>.

D. J. Kirkland, *Basic Mutagenicity Tests: UKEMS Recommended Procedures*, Cambridge University Press, Cambridge, 2005.

2.8 Carcinogenicity

carcinogenicity
Ability of an agent to induce malignant neoplasms, and thus cancer.

Cancer is a general term for a group of diseases characterized by the presence of dedifferentiated cells with a tendency to uncontrolled growth and invasion of other tissues. Uncontrolled growth of cells produces a swelling or a lump, usually referred to as a tumour. Tumours that grow slowly, are non-invasive, and are encapsulated are said to be benign, while tumours releasing invasive cells are said to be malignant.

Tumours may be described as 'neoplasms'. Neoplasms are the result of neoplasia, new cell division leading to abnormal arrangement of cells that may result in the production of either benign or malignant tumours. However, some neoplasias do not result in tumours, *e.g.* neoplasias of cells in the blood system. Neoplasia must not be confused with hyperplasia, excessive multiplication of normal cells in the normal tissue arrangement. The term metaplasia is also used, and refers to the replacement of one differentiated cell phenotype with another (*e.g.* the replacement of squamous epithelia with columnar epithelia in Barrett's oesophagus, or of ciliated columnar epithelia with stratified squamous epithelia in the bronchus of a smoker). If the stimulus that leads to metaplasia persists, it may lead to neoplastic transformation.

Carcinogens, agents that cause cancer, may be chemical, physical or biological. Some of them are organ-selective in their action, *e.g.* 1,2-dimethylhydrazine has been associated particularly with colon cancer in rats. Chemical agents may be so-called ultimate carcinogens, active without metabolism, or procarcinogens (proximate carcinogens), which must be metabolized into carcinogenic derivatives. Physical agents include ionizing radiation (which may be produced by the decay of radionuclides) and UV light. Biological agents include some types of bacteria and viruses. In general, anything that can cause chronic irritation and inflammation has the potential to be carcinogenic. Such agents include insoluble particulates. This property may not be absolutely dependent on the chemical nature of the particulates but may be, at least partly, a result of the physicochemical properties of surfaces. This may explain why chemically dissimilar particulates such as asbestos and crystalline silica can cause lung cancer.

Mechanism of Carcinogenesis. Carcinogenesis is a multistage process. It is not related in any simple way to acute toxicity. Seldom does a single exposure to a potential carcinogen, even in a very large dose, cause cancer. Chronic exposure is usually required. Removal from exposure may permit recovery, *e.g.* repair of damaged DNA that could otherwise lead to cancer. The development of cancer probably involves a minimum of three stages: initiation, promotion and malignant conversion (Figure 16). Initiation is the first stage in which a

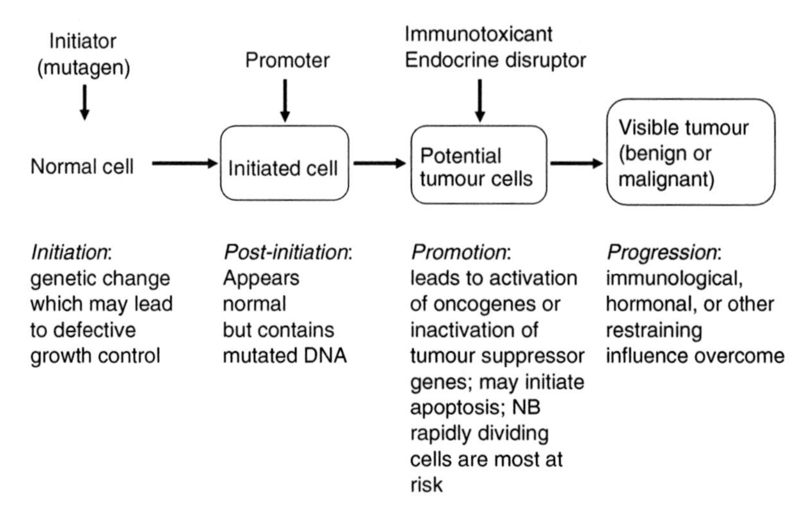

Figure 16 Stages of carcinogenicity following genetic change (mutation).

single somatic cell undergoes non-lethal, but heritable, mutation. Unlike its neighbours, the initiated cell can escape cell regulatory mechanisms restricting cell division.

Promotion, the second stage, occurs when the initiated cell is exposed to a tumour promoter, a substance that causes initiated cells to undergo clonal expansion. The signal to expand clonally may be the result of a direct effect of the tumour promoter on the initiated cell or of an indirect effect on the adjacent cells altering their interaction with the initiated cell. Tumour initiation and promotion together produce a relatively benign clonal expansion. An example of a promoter is the drug phenobarbital, which promotes liver tumours.

Malignant conversion is the third stage of carcinogenesis. This process is generally slow, occurs over a long time and is affected greatly by agents that alter growth rates (*e.g.* hormones, growth factors, vitamin A and the retinoids, vitamin D, folate and calcium). It is thought to be caused by repeated rounds of division and further chromosome damage, in addition to that responsible for initiation. Progression to malignancy is probably the most complex of the three stages. Both acquired genetic and phenotypic changes occur, and cell division becomes rapid. Retinoids affect this stage, as do other inhibitors of growth (*e.g.* polyamine synthesis inhibitors) and antioxidants. As the tumour progresses, sensitivity to dietary compounds, inhibitors of growth and enhancers of differentiation gradually disappears until the tumour becomes autonomous and controllable only by drastic intervention.

Genotoxic and Non-genotoxic (Epigenetic) Carcinogens. Genotoxic carcinogens are agents that either directly or indirectly through their metabolites act directly on the DNA of cells. Genetic change (mutation) results either from chemical modification of the nucleotide sequence, from the introduction of errors when the DNA is transcribed from chromosomes that have been broken during mitosis. There is postulated to be no threshold dose for genotoxic effects. In other words, it is supposed that even one molecule of a genotoxic carcinogen may react with DNA to cause a mutation that may lead to cancer.

Non-genotoxic (epigenetic) carcinogens do not modify the DNA nucleotide sequence causing mutations but act by other mechanisms, *e.g.* switching on cell division, enzyme stimulation or inhibition (which may modify DNA by adding groups to it, *e.g.* methyl groups, or removing them), immunomodulation, hormone balance changes or a combination of these effects (Section 2.5). Subsequent to these changes, mutations may occur, reinforcing the malignant properties of resultant tumour cells. A threshold has been demonstrated for such non-genotoxic epigenetic effects. Ultimately, epigenetic mechanisms lead to the formation of malignant cells that contain mutations. However, the initial events in the process are not caused by mutation and this is why a threshold is observed. Epigenetic agents operate largely as promoters of cancer and usually require high and sustained exposures. For example, dietary fat can act as an enhancer of cancer induction at about 40% of total calories available, probably by enhancing the availability of fat-soluble primary carcinogens. The effects of epigenetic agents, unlike those of genotoxic agents, are generally reversible. Accordingly, distinction between these two types of carcinogens is critical for informed risk assessment.

An important example of an epigenetic mechanism of carcinogenicity that has had consequences for carcinogenicity classification is peroxisome proliferation, recently reviewed by Melnick, Thayer and Bucher (*see* Further Reading). Peroxisomes are subcellular structures that contain several oxidase enzymes. Agents that cause increases in their numbers are called peroxisome proliferators. Many peroxisome proliferators are rodent carcinogens, and the mode of action proposed for rodent liver tumour induction by peroxisome proliferators involves activation of the peroxisome proliferator-activated receptor (PPARα), which results in altered transcription rates of genes that regulate cell proliferation and apoptosis. However, this hypothesis has not been tested with experimental studies demonstrating consistent increases in liver tumour incidence as a direct function of the time-dependent induction of cell proliferation and suppression of apoptosis in rats and mice treated with peroxisome proliferators. A further problem with the hypothesis is that increases in cell proliferation are generally only a transient response that returns to control levels within about 2–4 weeks after initiation of continuous exposure, whereas tumour induction requires chronic exposure for most peroxisome proliferators. Peroxisome proliferation *per se* does not appear to be a causal event in liver carcinogenesis.

Despite the above observations, there has been a general acceptance of the hypothesis that di(2-ethylhexyl)phthalate (DEHP) induces liver tumours in rats and mice by a non-DNA-reactive mechanism involving peroxisome proliferation. This mechanism is considered not to be relevant to humans because peroxisome proliferation has not been documented either in human hepatocyte cultures exposed to DEHP or in the liver of exposed nonhuman primates. Consequently, IARC downgraded the classification of DEHP from 'possibly' to 'not classifiable as to its carcinogenicity to humans'. Altered expression of cell growth and apoptosis genes by DEHP has not been demonstrated to be dependent on PPARα activation. In fact, the mechanism of tumour induction by DEHP is not known.

Available mechanistic data do not support the hypothesis that liver tumour induction in rats and mice occurs by a mechanism involving peroxisome proliferation that is not relevant to humans. A recent study has shown that dietary administration of DEHP induces liver tumours in mice lacking a functional PPARα gene. This finding emphasizes the need to test mechanistic hypotheses that, if relied on, might lead to erroneous cancer risk classifications and inadequately protective public health decisions.

Carcinogenicity Testing. The operational definition of a carcinogen is an agent having the ability to cause:

1. an increased incidence of tumours over the incidence observed in controls;
2. an occurrence of tumours earlier than seen in controls;
3. the development of tumour types different from those normally seen;
4. an increased multiplicity of tumours in individual animals.

The conduct of carcinogenicity studies is very expensive in that:

1. large numbers of animals are required;
2. the duration should be almost the life-span of the animal (18 months for mice, 24 months for rats);
3. animals need to be euthanized at intervals with particular attention being paid at necropsy to the morphological examination of tumours and the histological identification of cell hyperplasia, preneoplastic nodules, and whether tumours are benign or malignant.

Since carcinogenesis is a multistep process occurring over a prolonged period of time and progressing through several stages, only a few rodent carcinogenicity studies permit the study of mechanisms of action. The end-point(s) of these studies is to consider one or more of the criteria set out above for a carcinogen without considering how and why the agent acts as it does.

Purely mechanistic studies can be conducted *in vivo* using special substrains of rodents having a predisposition toward developing high incidences of specific tumour types. These susceptible substrains of mice and (or) rats are already compromised genetically toward certain types of tumours (pulmonary, dermal, hepatic, breast, *etc.*), and chemical exposure for 90–120 days frequently causes an increased incidence and (or) earlier appearance of tumours than seen in control animals of the substrain (these latter having a higher incidence than seen in other strains). Most 'normal' rodent strains will show a 1.0–5.0%

incidence rate of tumours in the untreated state. The special substrains mentioned above may have a 'normal' control incidence rate of 30% that, following treatment with an appropriate carcinogen, may be as high as 60–90%.

The International Agency for Research on Cancer (IARC). Identifying carcinogens is essential for their regulation and for reduction of the incidence of cancer in populations at risk. This is the main activity of IARC. IARC convenes expert groups that assess the available evidence and, on the basis of this evidence, produces a classification of suspected carcinogens. The evidence is subsequently published with the rationale of the classification in IARC Monographs.

Although IARC classification of carcinogens, when available, is generally given precedence, it is not necessarily always accepted. Thus, there are subtly different classification schemes applied by the European Union, the US EPA and the American Conference of Governmental Industrial Hygienists to give only three examples. These may be found described in detail in the appropriate further reading cited below. A concise comparison may be found in Annex 3 of J. H. Duffus. M. Nordberg and D. M. Templeton, Glossary of terms used in toxicology, 2nd edn, *Pure Appl. Chem.*, 2007, **79**(7), 1153. Available at <http://media.iupac.org/publications/pac/2007/pdf/7907x1153.pdf>

Further Reading

ACGIH, *2009 Guide to Occupational Exposure Values*, ACGIH, Cincinnati, 2009.

Cancer Information Service (U.S. National Cancer Institute). Available at <http://cis.nci.nih.gov/>.

International Agency for Research on Cancer, Preamble to the IARC Monographs (amended January 2006). Available at <http://monographs.iarc.fr/ENG/Preamble/index.php>.

International Agency for Research on Cancer, Monographs available in PDF format, 2008. Available at <http://monographs.iarc.fr/ENG/Monographs/PDFs/index.php>.

Y. Ito, O. Yamanoshita, N. Asaeda, Y. Tagawa, C. H. Lee, T. Aoyama, G. Ichihara, K. Furuhashi, M. Kamijima, F. J. Gonzalez and T. Nakajima, *J. Occup. Health*, 2007, **49**, 172.

R. Melnick, K. A. Thayer and J. R. Bucher, *Environ. Health Perspect.*, 2008, **116**, 130.

USEPA, *Guidelines for Carcinogen Risk Assessment (2005)*, U.S. Environmental Protection Agency, Washington, D.C., 2005. Available at <http://cfpub.epa.gov/ncea/cfm/recordisplay.cfm?deid=116283>.

R. A. Weinberg, *The Biology of Cancer*, Garland Science, London, 2006.

European Commission. REGULATION (EC) No 1272/2008 OF THE EUROPEAN PARLIAMENT AND OF THE COUNCIL of 16 December 2008 on classification, labelling and packaging of substances and mixtures, amending and repealing Directives 67/548/EEC and 1999/45/EC, and amending Regulation (EC) No 1907/2006, *Official J. Eur. Union*, L 353/1, 16 December 2008. Available at <http://eur-lex.europa.eu/LexUriServ/LexUriServ.do?uri=OJ:L:2008:353:0001:1355:EN:PDF>.

2.9 Pharmacogenetics and Toxicogenetics

pharmacogenetics
Study of the influence of hereditary factors on the effects of drugs on individual organisms.
toxicogenetics
Study of the influence of hereditary factors on the effects of potentially toxic substances on individual organisms.

The term pharmacogenetics acknowledges the role of genetic factors in pharmacology, especially in drug metabolism. In the 1950s it was recognized that certain individuals (about 1 in 3500 Caucasians) had an impaired ability to hydrolyse the muscle relaxant succinylcholine, leading to prolonged muscle paralysis after anaesthesia. This was shown to be due to a genetic variant of the phase I enzyme butyrylcholinesterase, and thus was pharmacogenetics born. Other examples soon followed. One of the best-known examples of genetic diversity of drug metabolism is variation in isoforms of the phase II enzyme *N*-acetyltransferase that affects the rate at which certain drugs such as hydralazine, isoniazid and procainamide are acetylated. Based on differences in half-lives of these drugs, the population is divided into 'slow acetylators' and 'fast acetylators'. Another example is provided by thiopurine methyltransferase, which metabolizes 6-mercaptopurine and azathiopurine, drugs used in treating leukaemia. In thiopurine methyltransferase deficiency, metabolism by alternative pathways produces a metabolite that is toxic to the bone marrow and potentially leads to bone marrow suppression. In most affected people, this deficiency results from one of three variant alleles. One in 300 people have two variant alleles and need only 6–10% of the standard dose of the drug. They are at risk of developing severe bone marrow suppression.

The cytochrome P450 enzymes are an extremely important group in phase I metabolism and their genetics have been widely studied. One of the best examples is CYP2D6. Many drugs as diverse as codeine, metoprolol and dextromethorphan are metabolized by CYP2D6. Some 5–10% of Caucasians show decreased metabolism of the antihypertensive debrisoquin, with increased ratios of parent drug to 4-hydroxydebrisoquin in the urine as a consequence of genetic variation in the CYP2D6 gene that leads to a less active or inactive cytochrome.

They have an exaggerated response to debrisoquine, resulting in hypotension. Bearing in mind the large number of drugs metabolized by CYP2D6, such patients are classed as 'poor metabolizers'. In contrast, some people have multiple copies of the CYP2D6 gene. They have an inadequate response to standard doses of drugs metabolized by CYP2D6, and are termed 'extensive metabolizers'. Relatively infrequent in Northern Europe, the occurrence of multiple copies of the CYP2D6 gene is as high as 29% in East African populations.

Allelic variants are now known that affect virtually all classes of drugs. These not only affect metabolizing enzymes but also drug targets, as well as transporters that determine absorption, distribution and excretion of drugs. Single nucleotide polymorphisms (SNPs) are defined as allelic variants with a frequency of greater than 1%. There are approximately ten million SNPs in the human genome. Association studies linking genotype with drug-response phenotype are now being considered on a broad scale. Drug efficacy is, of course, affected by metabolism, and it is hoped that, by understanding individual differences in metabolic profiles, therapy can be targeted more effectively to the individual. Adverse and idiosyncratic drug reactions, too, often result from genetic variation in drug metabolism.

Extending the discussion beyond drugs, the toxicity of any substance often depends on its disposition and metabolism, and hence on genetic variations in the responsible carriers, transporters and enzymes. In the past, occupational exposure to benzidine occurred, and slow acetylators were at greater risk of developing benzidine-related cancer of the bladder. Polymorphisms in CYP1A1 (aromatic hydrocarbon hydroxylase) are associated with variable risk of cancer caused by polyaromatic hydrocarbons. The glutathione *S*-transferase allele GSTM1 has been associated with lung cancer, *e.g.* in exposure to cigarette smoke, and skin cancer caused by arsenic in drinking water. NAT1 and NAT2 alleles of *N*-acetyltransferase are involved in metabolizing aromatic and heterocyclic aromatic amines, and have been associated with lung cancer in smokers and also colorectal cancer.

The terms pharmacogenomics and toxicogenomics refer to the study of the genome as it relates to pharmacology and toxicology. That is, they refer to the application of the science of genomics to understanding differences in toxicity and drug metabolism. Notably, however, in common usage a clear distinction between, for example, toxicogenetics and toxicogenomics is not always made. Apart from understanding individual susceptibilities, the techniques of toxicogenomics promise new mechanistic insights into mechanisms of toxicity. Using microarrays to study gene expression at the RNA level can reveal early changes that precede clinical or gross histopathological changes. This in turn could lead to development of early biomarkers of toxic injury. With more prolonged exposure, adaptive changes may be expected in clusters of genes that represent signatures characteristic of certain pathways of toxicity. An interesting newer application of toxicogenomics is to assessing validity of extrapolation from test organisms to man. The genomic sequences of humans and, for example, mice are compared and conserved regions identified. These are used to construct microarrays of orthologous[i] genes for the two species. Measurement of gene expression against these microarrays in both species following exposure to a toxic substance allows assessment of conserved toxicological end-points at a molecular level.

Finally, it will be realized that there are ethical issues in determining individual differences in susceptibility to toxic substances. For example, if an individual had an increased susceptibility to a toxic effect from an exposure likely to occur in a particular work place, the employer might decide to limit employment, independently of the worker's acceptance of the risk.

Further Reading

E. Hodgson and E. L. Croom, Phase I – toxicogenetics, in *Molecular and Biochemical Toxicology*, John Wiley & Sons, Inc., Hoboken, 2008, p. 205.

Y. Tsuji, Regulation and polymorphisms in phase II genes, in *Molecular and Biochemical Toxicology*, John Wiley & Sons, Inc., Hoboken, 2008. p. 239.

2.10 Epigenetics

epigen/esis n., **-etic** adj.
Phenotypic change in an organism brought about by alteration in the expression of genetic information without any change in the genomic sequence itself.

Note: Common examples include changes in nucleotide base methylation and changes in histone acetylation. Changes of this type may become heritable.

Until recently it has been dogma that gene expression is dependent on the primary DNA sequence (the genetic code), with regulation of gene transcription being determined by transcription factors or repressors whose transcription is in turn regulated in a tissue-dependent series of feedback mechanisms. In this paradigm, inherited genetic diseases arise from alterations in the primary DNA sequences that give rise to one or more proteins with consequently altered

[i] *Note*: 'Orthologous' is a term applied to genes (in different genomes) that are derived from a common ancestral sequence (homologous) by phylogenetic descent. It may also be applied to proteins or other gene products encoded by such genes. A closely related term is 'paralogous' that is applied to genes or nucleotide sequences within a single genome that arose from a single ancestral sequence by duplication and subsequently evolved independently. Again, this term may be applied to proteins or other gene products encoded by such a gene, or by the same gene in different individuals.

amino acid sequences. Epigenetics acknowledges that numerous factors outside the DNA sequence can affect the pattern of gene expression, that these alter phenotype and may cause disease, and that they can be heritable.

In earlier usage, epigenetics referred to any change in phenotype that was not accounted for specifically by changes in DNA sequence. In cancer biology, for instance, carcinogens that induce specific DNA mutations are considered to have no threshold effect for induction of cancer, whereas other (*i.e.* epigenetic) exposures may have a level below which excess cancers are not produced. Thus, carcinogens are sometimes divided into genotoxic and epigenetic carcinogens. Mechanisms of epigenetic changes in cancer can include silencing of tumour-suppressor genes, activation of oncogenes and defects in DNA repair. They can also include effects of substances with broad actions that modify the immune or endocrine systems. It has been practice to distinguish cancer initiators from promoters, and promoters have been considered to exemplify epigenetic mechanisms. These considerations would lead to a very broad definition of epigenetics. Today, the term is acquiring the more specific meaning of changes in DNA or histone structure, other than base sequence changes, that alter gene expression.

The human chromosome does not consist of elongated pieces of naked DNA awaiting the transcriptional machinery, but rather of DNA coiled and packed in a highly organized manner, notably by interactions with various proteins known as histones. The winding of DNA around successive multimeric histone structures gives rise to a 'beads-on-a-string' type of structure. Each individual histone/DNA 'bead' is referred to as a nucleosome and the 'string' is chromatin. To be transcribed, a given DNA sequence must partially unwind or dissociate from the histone core. A higher level of structure is determined by interactions among nucleosomes that may give rise to tighter or looser regions of chromatin packing along the chromosome, or expose regions between nucleosomal or chromatin domains to binding of repressor proteins that influence structure and transcription of extended regions of chromatin. In general, a more open chromatin structure, referred to as euchromatin, is active (more conducive to gene expression) while more condensed regions called heterochromatin are silent (suppressing expression). From steric arguments, it is logical that accessibility to the transcriptional machinery depends both on loosening of the heterochromatin structure and unwinding of DNA within the nucleosome. Chemical modifications of both DNA and histones influence the degree of compaction and winding or unwinding.

Various modifications of histones are documented, including acetylation, methylation and phosphorylation. Ubiquitination also occurs and targets the histone protein for degradation by the proteasome. It has been proposed that the pattern of N-terminal modifications of histones represents a new kind of genetic code that determines gene expression by changing histone conformation. The enzymology of histone acetylation has been widely studied. The extent of acetylation depends on the balance of activity between histone acetylases and histone deacetylase. Acetylation of lysine residues in the N-termini of histones H3 and H4 masks positive charges that otherwise interact with acidic residues in H2 histones of adjacent nucleosomes. This leads to a more open chromatin structure and favours gene expression. The histone acetylase/deacetylase system is of great importance in toxicology and is visited again in the discussion of mutagenicity.

The major epigenetic chemical modification of DNA is methylation. Most important is the relative hyper- or hypomethylation of position 5 of cytosine in CpG dinucleotides that tend to cluster in promoter regions of genes. The degree of methylation of a stretch of DNA is generally inversely correlated with its degree of expression. Increased methyl-CpG content is associated with increased heterochromatin formation. That hypermethylation tends to occur in promoter regions suggests it may interfere with transcription factor binding, and there is evidence that some transcriptional repressors bind with increased affinity to methylated CpG sequences. However, it is not yet clear whether methylation of DNA is directly responsible for suppression of transcription or whether it acts in conjunction with histone tagging or through influencing nucleosomal structure. Furthermore, there is evidence that DNA methylation and histone deacetylation may be linked. A protein MECP2 has been identified that binds to methylated CpG sequences and recruits histone deacetylase, amplifying the transcriptional suppression.

DNA methylation participates in other aspects of epigenetic control. Imprinted genes are those in which one allele (either the maternal or paternal) is effectively silenced in early development, and this usually involves hypermethylation of the silent allele. In addition, certain DNA sequences called transposons are prone to self-cloning themselves into other regions of the chromosome, where they may either disable or hyperactivate a target gene. Methylation tends to suppress the mobility of transposons.

The dogma that only DNA that ultimately codes for a protein represents a gene has also been upset by the demonstration of numerous RNA-only genes. (Some purists who prefer to reserve the word 'gene' for a DNA sequence that is expressed as protein also prefer the term 'transcriptional unit' for any DNA sequence that is actively transcribed.) These are genes that encode RNA that is not translated, but affects the expression of other genes nevertheless. Numerous examples are known where the complementary strand is transcribed as an antisense RNA that blocks translation of its sense partner. Other genes code short hairpin RNAs that silence other genes through RNA interference. The involvement of untranslated RNA sequences in epigenetic phenomena is an active area of investigation.

Various definitions have indicated that the epigenetic phenomenon must persist for at least one generation, or be heritable. However, if heritability is taken as part of the definition of epigenetics in current usage, then it is necessary to demonstrate inheritance in the F3 generation. This is because an exposure that affected (injured) the F1 daughter *in utero*, for example, could also affect her eggs and thus have a direct but not heritable affect on the F2 generation. Thus, true demonstration of heritability is not always easy to achieve. When it occurs, it is reminiscent of Lamarckianism, the once discredited view that characteristics acquired from environmental exposures during an individual's lifetime could be passed on to progeny.

How are these patterns of chemical modification passed on from cell to cell or generation to generation independent of the genetic code? The answers are complex and not well understood. An early example was the inheritance of the trait *callipyge* (Greek, beautiful buttocks) in sheep developing advantageous hindquarters. A decade of experimentation revealed the unusual inheritance pattern of this trait is due to a conserved protein-coding gene and RNA-only genes on the same chromosome, that are transcriptionally regulated epigenetically in response to a G-to-A base change in a region of 'junk DNA' 30 000 bases away from the nearest known gene. Loss of imprinting (*e.g.* of the insulin-like growth factor IGF2) has been associated with some sporadic cases of colon cancer and several rare genetic diseases. Identical twins discordant for development of Beckwith-Weidemann syndrome were found to differ with respect to imprinting on a region of chromosome 11, the affected twin having failed to imprint.

Epigenetic phenomena are beginning to take a prominent place in toxicology. Their involvement in nickel and arsenic toxicity are good examples. Although divalent nickel cations are at best weakly mutagenic, nickel compounds have been classified by IARC as human carcinogens, and there is mounting evidence that any such carcinogenic effects may be through heritable changes in DNA methylation and/or histone acetylation. In the nucleus, Ni^{2+} competes with Mg^{2+} and binds selectively to heterochromatin, probably through specific interactions with various histone sites. Resulting decondensation and damage to heterochromatin correlates with morphologic transformation. Damage includes heritable chromosome deletions and hypermethylation of a senescence gene on the X-chromosome. Silencing of the tumour suppressor gene p16 has also been demonstrated following exposure to various nickel compounds.

A unique epigenetic mechanism is also suggested for toxicity of carcinogenic arsenic compounds. Arsenic in divalent and pentavalent states is detoxified through mechanisms that involve formation of methylated derivatives. The methyl donor is *S*-adenosylmethionine, the same donor involved in DNA methylation. Exposure to some arsenicals leads to depletion of *S*-adenosylmethionine, and consequently to global DNA hypomethylation. One idea is that hypomethylation of proto-oncogene may lead to their overexpression and ultimately to cell transformation. This also illustrates the point that both DNA hypermethylation and hypomethylation may lead to cancer, depending on whether the targets are pro- or anti-carcinogenic genes.

The above examples reflect a recent focus on toxicological aspects of epigenetics relating to cancer. The idea that cancer can be explained by single or multiple mutational hits on genes that stimulate or inhibit proliferation is being reconsidered. Clearly, heritable effects that affect gene expression give rise to at least some cancers, independent of any alterations in the DNA sequence of the gene itself, and this adds a new layer of complexity to our attempts to understand malignant transformation. However, because these epigenetic phenomena involve dynamic chemical modifications to the genome that may switch on or off during the life of the cell, they present potentially more tractable targets for therapeutic intervention than changes in the primary DNA sequence itself.

Finally, notably, the term epigenomics has come into prominent use. This refers to the study of the epigenetic factors in illness or susceptibility using a genomics approach (Section 2.11). While the term epigenetics refers to changes resulting from one or more individual genes, epigenomics refers to a more global analysis of epigenetic effects across the genome. Thus, it relies on techniques that give a genome-wide picture of epigenetic phenomena, for instance using high-throughput microarray technology. Currently these approaches are best developed for DNA methylation, and for histone modifications using DNA crosslinking followed by chromatin immunoprecipitation and microarray (ChIP-on-chip) analysis. Epigenetic control of stem cell differentiation and organ development may underlie a broad spectrum of disease. Initially, studies are being concentrated on cancer and aging. There is now considerable hope that insight into major psychiatric illnesses like schizophrenia and bipolar disorder will be found through epigenomics research.

Further Reading

C. Ptak and A. Petronis, Epigenetics and complex disease: from etiology to new therapeutics, *Annu. Rev. Pharmacol. Toxicol.*, 2008, **48**, 257.

2.11 Genomics, Proteomics and Related Terms

genomics
1 Science of using DNA and RNA based technologies to demonstrate alterations in gene expression.
2 (in toxicology) Method providing information on the consequences for gene expression of interactions of the organism with environmental stress, xenobiotics, *etc.*
proteomics
Global analysis of gene expression using various techniques to identify and characterize proteins.

Note: It can be used to study changes caused by exposure to chemicals and to determine if changes in mRNA expression correlate with changes in protein expression: the analysis may also show changes in post-translational modification, which cannot be distinguished by mRNA analysis alone.

The publication of the first drafts of the complete human genome in the journals *Nature* and *Science* in 2001 heralded a new age in biology, and a proliferation of terminology – some good and some bad – that accompanied the new way of thinking. It is now believed that we can know the complete set of genes for a given organism – its genome. If that is so, presumably it should be possible to know its complete set of proteins, carbohydrates, metal-binding species and even all the possible phenotypic traits expressed by the organism with the same certainty. Some of the methods and successes of genomics and proteomics will be discussed here. Problematic issues relating to other comprehensive '-omics' will be reviewed at the end of this section.

The human genome project, promulgated in the 1990s, finished ahead of schedule and revealed that the human species has fewer genes than expected. While prior estimates had ranged to upwards of 100 000, we now know that about 23 000 DNA sequences (genes) code for the proteins that contribute to our phenotype. This number does not correlate well with what we understand as biological complexity. We have fewer coding genes than the rice plant, and the minute worm *Caenorhabditis elegans* has more than the fly *Drosophila melanogaster*. However, we have come to realize that the total picture of an organism includes genes that code only for RNA that is not translated into protein, and a host of epigenetic phenomena (Section 2.5). Clearly, cataloguing the set of coding DNA genes for a given species is a long way from understanding the complete phenotype of that species, but nevertheless provides a wealth of information to approach the biology and pathology of the species. We can now compare the generic human genome with the nearly complete sequences of James Watson's and Craig Ventner's DNA, and from this we learn that inter-individual differences affect about 3.3 million base pairs considering only single base pair changes, but when insertions and deletions are included this number rises to around 15 million. With the draft human genome behind us, other genomes that have been tackled early on include those of model organisms for biological research (*Drosophila*, *C. elegans*, *Arabidopsis thaliana*, *Danio rerio*, the mouse and various yeasts) and those of agricultural significance, including rice and maize. Even extinct species such as the mammoth are yielding their genetic identity, and may one day be reconstructed. These successes have evolved hand in hand with technologies for large-scale sequencing and bioinformatic techniques for sorting, assimilating and storing sequence data.

Activities referred to as genomics today include not only the automated pursuit of more (and more complete) genome sequences but also the use of surveying techniques for gene expression under various circumstances. These survey approaches commonly include microarray technology, where nucleic acid sequences, typically either primary DNA sequences or complementary cDNA sequences, are immobilized on a support. In practice, several thousand sequences from an organism can now be immobilized on a patch of a glass microscope slide. Total RNA being translated by a given cell or tissue preparation is tagged for detection, usually with fluorophores, and then hybridized against the target array and detected by scanning laser fluorimetry. Quantitation of the fluorescent signal on each spot gives a snapshot of the expression of genes in the cell or tissue at the moment it was harvested. An exciting application of these genomic and proteomic approaches is the possibility of high-throughput screening of substances for a mechanistic understanding of toxic effects that could bypass the need for animal studies. Additionally, the full range of toxic potential of a substance could be gained from a few animals, without sacrificing individual animals for specific end-points.

Three considerations are worth mentioning in the interpretation of such experiments: (i) *Description of the experiment*: It was recognized early in the microarray era that replication of experiments between laboratories was necessary to validate results that are so prone to artefact, and this requires careful description of the experimental protocol. An international consortium of journal editors devised the 'minimum information about a microarray experiment' (MIAME) protocol that should be provided to contribute the results of a microarray analysis to the literature. The protocol requires details of (a) experimental design, (b) array design, (c) sample selection, (d) hybridization protocol, (e) image analysis and (f) normalization and controls for comparison. It is optimistic to think that this information will be sufficient to consolidate data among many laboratories, but it is a starting point.

(ii) *Statistical analysis*: Because there are many levels of potential analytical variability, there is much discussion of how many replicates of a microarray experiments should be performed and how statistical validity should be determined. Costs still frequently limit the number of replicates, at least in the academic research laboratory. Normalization of the signal intensity is a necessary precursor to meaningful analysis, and data processing through available routines such as significance analysis of microarrays (SAM) and nonlinear regression by the LOESS method is mandatory to minimize artefact.

(iii) *Data management*: Because one microarray experiment can potentially generate meaningful data on the expression of thousands of genes, data management is a consideration. It is the rule, rather than the exception, that a well-considered microarray experiment provides a research laboratory with enough information to fuel months or years of ancillary confirmation and intriguing follow-up.

Another powerful aspect of genomics is phylogenetic analysis. This uses analysis of gene sequences to study and establish evolutionary relationships between and among organisms. Gene trees are built up, sometimes referred to as 'cladistics' (from the Greek *clados*, a branch, referring to descent from a single ancestor). The end product from cladistic analysis is a phylogenetic tree that represents the evolutionary proximity of organisms based on similarities in the sequences of one or more genes.

Proteomics is the next logical step after definition of the genome. One attempts to identify all the proteins expressed by the genome, and while in principle this might be achieved by *in silico* translation of all the genes, this is not useful because we still cannot predict cell- and tissue-specific controls on transcription, mechanisms of alternative splicing, or epigenetic controls on gene expression/modification. Therefore, we need to measure the protein complement expressed at a given time in a given cell or tissue to understand the manifestation of the genome in a given state of health, disease or an environmental situation. The standard approach here is two-dimensional electrophoresis followed by mass spectrometry. After careful isolation of all proteins in a sample, in the presence of broad inhibitors of proteases, phosphatases and other degradative enzymes as the desired result dictates, the solubilized protein sample is first separated by isoelectric focusing on an ampholine gradient, which separates proteins based upon their isoelectric points. This partially separated mixture is then subjected to electrophoresis in the second dimension on a polyacrylamide gel, based on differences in size and charge. Individual proteins are recognized non-quantitatively and non-specifically, by stains based on Coomassie blue dye or much more sensitive silver staining. This approach is currently able to resolve a few thousand proteins in a sample. Spots are then cut from the gel, eluted and subjected to analysis by mass spectrometric techniques.

The genome is well defined. The proteome is approachable by current methods. After this, we must exercise linguistic restraint. There is a trend to classify other collections of molecules as if in a complete descriptive context, and thus emerge terms like kinome (the set of expressed kinases), glycome (the complement of synthesized carbohydrates) and metallome (sometimes meant to refer to the set of expressed metalloproteins). For example, surveys of the kinome depend on immunoblotting techniques and thus are governed by the availability, specificity and affinity of antibodies. They are incomplete and non-quantitative; the practice is without guidance, and the nomenclature is imprecise. The origins of the term 'genome' are not helpful in constructing terms to describe these other sets of molecules. It is an amalgamation of gene (from the Greek *genea*, generation or race) and chromosome (again Greek, *chroma*, colour and *soma*, body, based on the staining properties observed by the early cell biologists). By loose analogy, various -ome and -omics words come into the language. IUPAC has already defined interactome, phenome and transcriptome. While they may serve some scientific purpose, they can also be problematic. As an example, the term metallome has been used to describe the set of molecular metal-binding species in a given organism. But should this refer to metal-binding proteins (a subset of the proteome) or to all metal-binding species, even including metal salts that may form with available counterions? This is a problem of chemical speciation, not of comprehensive description. The concept has even been further subdivided in coining terms such as zincome and selenome, to refer to the species that bind zinc or selenium. This practice seems to be both scientifically distorting and etymologically unfounded, and should surely be discouraged. Even the more accepted terms 'metabolomics' and 'metabonomics' are a case in point. Ambiguity in definition belies the fact that no complete set of metabolites can be defined in the same way that a genome can be sequenced, as these will depend on individual differences in enzyme activities, environmental exposures to xenobiotics, *etc*. However, notably, some biologists consider metabolomics to be the study of the collection of metabolites themselves, and take metabonomics to describe the study of the metabolic changes that cells undergo in response to stressors. The term epigenomics has a special status, and is mentioned in Section 2.10.

Further Reading

International Human Genome Sequencing Consortium, *Nature*, 2001, **409**, 860.

International Human Genome Sequencing Consortium, *Science*, 2001, **291**, 1145. Available at <http://www.genome.gov/>.

2.12 Apoptosis and Other Modes of Cell Death

apopto/sis n., **tic** adj.
Active process of programmed cell death, requiring metabolic energy, often characterized by fragmentation of DNA, and cell deletion without associated inflammation.
necro/sis n., /**tic** adj.
Sum of morphological changes resulting from cell death by lysis and (or) enzymatic degradation, usually accompanied by inflammation and affecting groups of cells in a tissue.

Several years ago it was recognized that cells may die by means other than a simple irreversible response to injury, ultimately leading to membrane rupture (*i.e.* necrosis), such as hepatic necrosis induced by trichloromethane or acute tubular necrosis in the kidney caused by mercury. The alternative appeared to follow a sequential program of events leading to a systematic disassembly of the cell without the release of proteolytic and proinflammatory contents. The process was termed apoptosis (Greek *apo*, from, and *ptosis*, to fall; hence, falling off) by Andrew H. Wyllie in 1972. There is no consensus as to whether the 'a' is long or soft, or the second 'p' silent, and several pronunciations are in common use.

Apoptosis. In contrast to necrosis, occurring in response to acute injury such as hypoxia or exposure to toxic substances, apoptosis is considered to occur in physiological circumstances such as atrophy or at certain stages of development, or in selected pathological circumstances. It is now also known to play a key role in balancing a stable cell population in a tissue in a triad of stem cell differentiation, cell proliferation and apoptotic cell loss. For many years the recognition of apoptosis was by visual criteria. Apoptosis involves individual cells, undergoing shrinkage and formation of apoptotic bodies, whereas in necrosis groups of cells are involved in swelling and tissue disruption. Phagocytosis of apoptotic bodies is in contrast to the inflammation and subsequent regeneration or fibrosis consequent upon necrosis. In apoptosis, organelles generally remain intact whereas necrosis is characterized by swelling of the mitochondria and endoplasmic reticulum. Especially characteristic changes are seen in the nucleus, where condensation of chromatin in apoptosis is accompanied by internucleosomal breaks (karyorrhexis) and a characteristic pattern of bands of multiples of approximately 200 base pairs on agarose gel electrophoresis ('DNA laddering'). In contrast, random DNA breaks precede loss of DNA (karyolysis) in necrosis.

Subsequently, a family of proteases was discovered that was responsible for the controlled cleavage of various substrates during the apoptotic program. These have in common an active centre cysteine thiolate that cleaves at aspartic residues. Called caspases (Cysteine ASPartate proteASES), these include upstream initiator caspases that cleave and activate downstream effector caspases, which in turn cleave other protein substrates. About 14 caspases are currently known in humans, and for a time caspase activation was a hallmark of apoptosis.

As more biochemical detail emerged, two biochemical pathways to apoptosis were distinguished, termed extrinsic and intrinsic. The extrinsic pathway responds to extracellular triggers such as tumour necrosis factor α and Fas ligand, which acts at so-called death-ligand receptors to activate caspase-8. The intrinsic pathway is mediated by destabilization of the mitochondrial membrane to release cytochrome *c*, which acts with caspase-9 to form the apoptosome. The pathways converge at the effector caspase-3. Both caspase-8 and the apoptosome cleave procaspase-3 to active caspase-3.

DNA damage is one trigger of the intrinsic pathway, with the tumour suppressor p53 mediating the signal between DNA damage and mitochondrial destabilization. The Bcl-2 proteins are a family of proteins that act at the mitochondrial membrane to either stabilize (anti-apoptotic, *e.g.* Bcl-2, Bcl-X_L) or destabilize (pro-apoptotic, *e.g.* Bax, Bak) the mitochondrial membrane. A key determinant of apoptosis is the balance in expression between pro- and anti-apoptotic Bcl-2 family proteins. Caspase-8 can cleave the Bcl-like protein Bid to a truncated form, tBid, which is pro-apoptotic, thus providing cross-talk between the extrinsic and mitochondrial-mediated (intrinsic) pathways.

Necrosis. Necrosis itself spans a range from disorganized destruction of the cell to a more highly organized or programmed necrosis. While it may arise as a result of a metabolic catastrophe, it may also progress through steps that involve signalling pathways, the active suppression of caspase activity and other apoptotic signals, and even require ATP. In 1999 Lemasters introduced the term 'necrapoptosis' to refer to a state poised between necrosis and apoptosis. This was an important recognition that cell death could not be described adequately by two discrete terms. It also placed an important focus on bioenergetics in cell death and survival, as explained further below.

Early in the third millennium we began to view cell death as a spectrum of mechanisms from apoptosis to necrosis. It was recognized that some cells that looked as if they were dying in apoptosis were not activating caspases, and thus the term caspase-independent apoptosis was introduced. It is still not precisely clear to what mechanism(s) this refers. One approach entertained for a short time was to refer to cell death as apoptosis, apoptosis-like, necrosis-like or necrosis. The term 'late apoptosis' has also been used, implying, perhaps, a bridging state between late apoptosis-like and early necrosis-like death, but these terms lack mechanistic definition.

Necrapoptosis. Necrapoptosis was considered to represent a state where the cell was poised between states of adequate energy supply and energy depletion. Its historical significance lies in this recognition of the role of bioenergetics in determining the pattern of cell death. However, we can now speak of ATP-independent apoptosis and ATP-dependent necrosis. Thus, while many processes of classical apoptosis (caspase cleavage, apoptosome formation, various translocation events) are ATP-dependent, ATP depletion can also trigger apoptosis. On the other hand, circumstances have been described where necrotic death cannot proceed without ATP hydrolysis to maintain expression of certain ion channels to allow transmembrane ion fluxes. It is also now established that ATP depletion, leading to increased ADP/ATP ratios, and ultimately increased concentration of AMP, activates AMP kinase. AMP kinase inhibits the target-of-rapamycin (TOR) protein, thus favouring death by autophagy. These bioenergetic events are not clearly understood at present.

Autophagy. Before we come to a contemporary, though surely-to-be short-lived classification of cell death, we should also consider the concept of autophagy. Upon nutrient deprivation, cells have the capacity to sequester their own contents in vacuoles, digest the contents and survive for a limited time by adaptive autodigestion to maintain ATP levels. Of course this can also become a mechanism of cell death. Interestingly, pan-caspase inhibition alone is sufficient to divert some cells to autophagy, suggesting a role of caspases in survival, and inhibition of autophagy (*e.g.* by disruption of Atg genes) can initiate apoptosis. Autophagy is now known to be under the control of at least 31 genes, referred to as Atg1–Atg31.

Classification of Cell Death Mechanisms. One fairly satisfactory contemporary classification of cell death mechanisms derives from terminology related to neuronal cell death, and considers five distinct alternatives on a biochemical basis, notably omitting necrosis:

1. Apoptosis. Type I cell death or nuclear cell death. Morphological hallmarks of apoptosis, with or without caspase activation. This is subdivided into caspase-dependent and caspase-independent apoptosis.
2. Autophagy. Type II cell death. Non-apoptotic and characterized by autophagic vacuoles, ultimately fusing with lysosomes to degrade vacuolar contents.
3. Programmed necrosis. Type III programmed cell death. Also called parapoptosis, or cytosolic death. Vacuolization, but no lysosomal involvement.
4. Non-apoptotic, non caspase-dependent death with nuclear shrinkage, dependent upon PARP (poly-ADP ribose polymerase) activation, PAR polymer accumulation and AIF (apoptosis-inducing factor) release from mitochondria. It has been called Parthanatos, and may be neuronal-specific.

5. A distinct Ca^{2+}-dependent mode of cell death – programmed, in that it is mediated by regulated activation of cathepsins and calpains, and is suppressed by calreticulin.

One of the most interesting challenges posed by this classification is to discern a mechanism for caspase-independent apoptosis that is distinctly apoptotic, yet also distinct from the other four mechanisms. Details of mechanisms 4 and 5 will be fascinating to learn.

Further Reading

K. L. Rock and H. Konoo, The inflammatory response to cell death, *Annu. Rev. Pathol. Mech. Dis.*, 2008, **3**, 99.

Concept Group 3. Concepts Applying to Whole Organism Toxicology

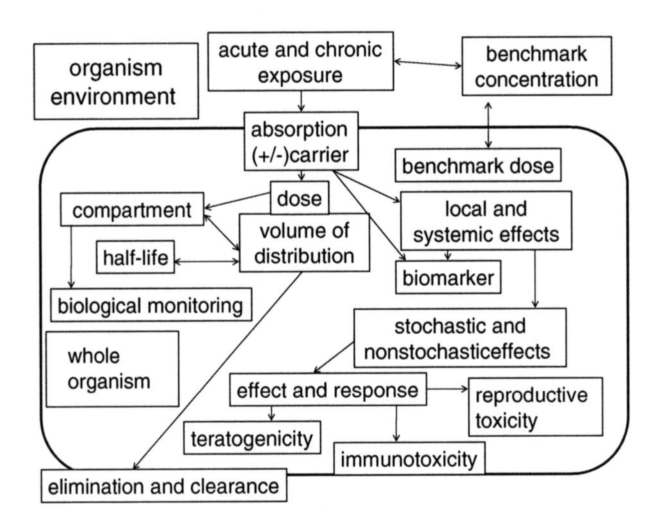

Concepts in Toxicology
By John H Duffus, Douglas M Templeton and Monica Nordberg

Published by the Royal Society of Chemistry, www.rsc.org

3.1 Benchmark Concentration and Dose

benchmark concentration (BMC)
Statistical lower confidence limit (CL) on the concentration that produces a defined response (called the benchmark response or BMR, usually 5 or 10%) for an adverse effect compared to background, defined as 0% (Figure 17).
benchmark dose (BMD)
Statistical lower CL on the dose that produces a defined response (called the benchmark response or BMR, usually 5 or 10%) of an adverse effect compared to background, defined as 0%.

The benchmark concept has been introduced in risk assessment to reduce the numbers of animals used in testing and to move away from compulsory LD_{50} determination. The aim is to be able to define a concentration between 'no observed adverse effect level' (NOAEL) and 'lowest observed adverse effect level' (LOAEL), which can substitute for them in risk assessment. In particular,

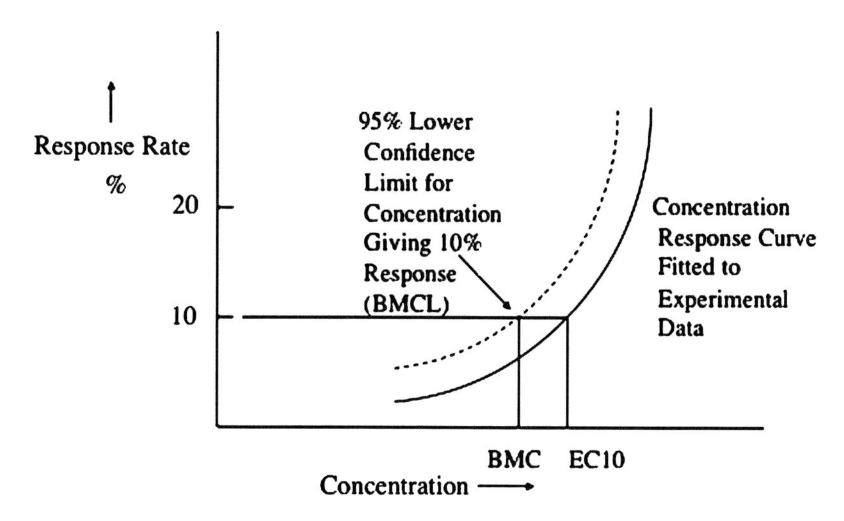

Figure 17 Dose–response curve showing BMC and BMCL (BMC 10% lower CL).

the benchmark concentration (BMC) or dose (BMD) is proposed as an alternative to the NOAEL. Using the NOAEL in determining acceptable human exposure values such as reference doses (RfDs) and reference concentrations (RfCs) has long been recognized as having limitations in that it:

1. is limited to one of the doses in the study and is dependent on study design;
2. does not take account of the variability in the estimate of the dose–response;
3. does not take account of the slope of the dose–response curve.

The NOAEL value is highly dependent on the quality of the data from which it is derived. The less precise these data are the larger the NOAEL value tends to become. This means that any derived permissible exposure level (PEL) may be too high for safety. Exposure to levels equal to (or even below) the NOAEL may still permit the occurrence of adverse health effects. This is why health-based recommended exposure limits are derived by dividing the NOAEL by an uncertainty factor to ensure adequate health protection. Such uncertainty factors are largely chosen *ad hoc* by the regulatory toxicologists involved.

Determination of the BMC or BMD requires quantitative analysis of the data relating level of exposure to the effects of a chemical on animal or human health. The aim of the data analysis is to determine, as accurately as possible, the relationship between a given exposure and the likelihood of its producing a defined harmful effect as measured by the response (the percentage of a test population showing the defined effect). The statistical uncertainty to which any data is invariably subject is incorporated into the calculations. The dose–response relationship obtained is plotted graphically and used to calculate the BMC or BMD: this is the concentration or dose that corresponds to a chosen statistical percentage likelihood of health impairment in the exposed population, for instance 5 or 10%. The BMC or BMD is divided by an uncertainty factor to yield a health-based recommended permissible exposure limit, chosen with the aim of protecting the whole population at risk. Software for calculating the BMD is available on the Internet (see below).

The BMC or BMD method takes account of research data uncertainties that are largely ignored in the NOAEL method, and, while the NOAEL is by definition the highest experimental dose applied that does not cause an adverse effect, the BMD is a quantity derived from all the available experimental values. The BMD method can also give information about the risks associated with exposure exceeding the health-based recommended exposure limits because it is related to a statistically modelled dose–response curve.

The BMC or BMD method requires the following three choices to be made:

1. the statistical likelihood of an effect occurring in the test population that would be used in the determination of the BMC or BMD;
2. the dividing line between an effect considered to be tolerable and one considered to be harmful;

3. the choice of a model function with which to describe the relationship between dose and response.

With current methodology, these choices have to be made and justified on a substance-by-substance basis. Derivation of uncertainty factors used to derive health-based permissible exposure limits is still *ad hoc*.

BMC and BMD refer to the central estimates, *e.g.* the effect concentration (EC_x) or the effect dose (ED_x) for dichotomous end-points (with x referring to some level of response above background, *e.g.* 5 or 10%). BMCL or BMDL refers to the corresponding lower limit of a one-sided 95% confidence interval on the BMC or BMD, respectively. This is consistent with the terminology used in the EPA's BMD software (BMDS) which is freely available on the Internet at <http://www.epa.gov/ncea/bmds.htm>.

Determination of Appropriate Studies and End-points on which to Base BMD Calculations. Following hazard characterization and selection of appropriate effect end-points for use in dose–response assessment, studies appropriate for modelling and BMC or BMD analysis should be evaluated. All studies that show a graded monotonic response with concentration or dose are likely to be useful for BMC or BMD analysis. The minimum dataset for calculating a BMD should show a significant dose-related trend in the selected effect end-point(s). It is preferable to have studies with one or more doses near the level of the benchmark response (BMR), usually 5 or 10%, to give a better estimate of the BMC or BMD and, thus, a smaller confidence interval. Studies in which all the concentration or dose levels show changes compared to the control values (unsuitable for NOAEL determination) can be used for BMC or BMD estimation, provided the lowest response level is reasonably close to the BMR.

There are at least three types of end-point data that may be available from toxicity testing: quantal (dichotomous), continuous or categorical. A quantal (dichotomous) response may be reported as either the presence or absence of an effect, a continuous response may be reported as an actual measurement, or as a comparison (absolute change from control or relative change from control). In the case of continuous data, the number of subjects, mean of the response variable, and a measure of response variability [*e.g.* standard deviation (SD), standard error (SE) or variance] are needed for each group. For categorical data, the responses in the treatment groups are often characterized for the severity of effect (*e.g.* mild, moderate or severe histological change).

Selection of end-points should not be limited to only the one with the lowest concentration causing the lowest-observed adverse effect, a kind of minimum value for the LOAEL. In general, end-points that have been judged to be appropriate and relevant to the exposure should be modelled if their LOAEL is up to ten-fold above the lowest LOAEL. This will help ensure that no end-points with the potential to have the lowest BMDL are excluded from the analysis based on the value of the LOAEL or NOAEL. Selected end-points from different studies that are likely to be used in the dose–response assessment should all be modelled, especially if different uncertainty factors may be used for different studies and end-points. As indicated above, the selection of the most appropriate BMCs (BMDs) and/or NOAELs (if some end-points cannot be modelled) to be used for determination of health-based PELs will be a matter of scientific judgement.

Selection of the BMR Value. Calculation of a BMD is directly determined by the selection of the BMR, the increase in the incidence of a given adverse effect in a population subjected to exposure to a toxicant. For quantal effects such as cancer or mortality, an excess incidence of 10% is usually the default BMR, since the 10% increase in detectable response in a given population is at or near the limit of sensitivity in most cancer bioassays and in some non-cancer bioassays as well. If a study has greater than usual sensitivity, then a lower BMR may be used, although the ED_{10} and LED_{10} should always be presented for comparison purposes.

For continuous data, if there is an accepted level of change in the end-point that is considered to be biologically significant, that amount of change should be selected as the BMR. Otherwise, if individual data are available and a decision can be made about what individual levels should be considered adverse, the data can be 'dichotomized' based on that cut-off value, and the BMR can be set as above for quantal data. Alternatively, not having any other idea of what level of response to consider adverse, a change in the mean equal to one control SD from the control mean can be used. The control SD can be computed including historical control data, but the control mean must be from data concurrent with the treatments being considered. Regardless of which method of defining the BMR is used for a continuous dataset, the effective dose corresponding to one control SD from the control mean response, as would be calculated for the latter definition, should always be presented for comparison purposes.

Choice of the Model to Use in Computing the BMD. The goal of the mathematical modelling in BMD computation is to fit a model to dose–response data that describes the dataset, especially at the lower end of the observable dose–response range. In practice, this involves first selecting a family or families of models for further consideration, based on characteristics of the data and experimental design, and fitting the models using one of a few established methods. Subsequently, a lower bound on dose is calculated at the BMR. USEPA guidance on BMC and BMD calculation recommends that 0.1 be used to compute the critical value for goodness of fit, instead of the more conventional values of 0.05 or 0.01, and that a graphical display of model fit be examined as well. For comparison of models and selection of the model to use for BMDL computation, the USEPA recommends the use of Akaike's information criterion (AIC), referenced in the Further Reading.

Computation of the Confidence Limit for the BMD (BMDL). The USEPA benchmark dose guidance document (*see* Further Reading) discusses the computation of the CL for the BMD, the fact that the method by which the CL is obtained is typically related to the data type, and the manner in which the BMD is estimated from the chosen model. Details for approaches to CL computationally specific to particular data types (quantal, clustered, continuous, multiple outcomes) are provided in the USEPA document.

Advantages of the Benchmark Dose (Concentration) Method. The advantages of using the BMD approach are many. Firstly, all the experimental data are used to construct the dose–response curve. Secondly, the variability and uncertainty are taken into account by incorporating SDs of means and, thirdly, the method represents a single methodology that can be applied to cancer and non-cancer end-points. It may also be possible to use fewer animals in testing.

Further Reading

H. Akaike, A new look at the statistical model identification, *IEEE Trans. Automatic Control* 1974, **19**, 716.

K. Z. Travis, I. Pate and Z. K. Welsh, The role of the benchmark dose in a regulatory context, *Regulatory Toxicol. Pharmacol.*, 2005, **43**, 280.

USEPA benchmark dose guidance document. Available at <http://cfpub.epa.gov/ncea/cfm/recordisplay.cfm?deid=20871>

3.2 Absorption

absorption

1 General process of one material (the absorbent) being retained by another (the absorbate); this may be the physical dissolution of a gas, liquid, or solid in a liquid, a gas or liquid in a solid, attachment of molecules of a gas, vapour, liquid, or dissolved substance to a solid surface by physical forces, *etc.* In spectrophotometry, absorption of light at characteristic wavelengths or bands of wavelengths is used to identify the chemical nature of molecules, atoms, or ions and to measure the concentrations of these species.

Note: This definition from the IUPAC 'Gold Book' requires that 'light' be interpreted as referring to all forms of electromagnetic radiation.

2 Of radiation. Phenomenon in which radiation transfers to matter which it traverses some or all of its energy.

3 In biology Penetration of a substance into an organism by various processes, some specialized, some involving expenditure of energy (active transport), some involving a carrier system, and others involving passive movement down an electrochemical gradient.

4 Uptake to the blood and transport *via* the blood of a substance to an organ or compartment in the body distant from the site of absorption.

Absorption of Radiation. Absorption of radiation is the prime consideration in radiation toxicology. Radiation can reach all tissues of the body directly from an external source, but the capacity to penetrate body tissues varies with the type of radiation. Radiation may be emitted as particles (α-, β- or neutron particles) or as high-energy electromagnetic waves such as X-rays or γ-radiation. Radiation may be ionizing or non-ionizing. Ionizing radiation is particle radiation in which an individual particle (*e.g.* a photon, an α-particle, or a β-particle) carries enough energy to ionize an atom or molecule (that is, to completely remove an electron from its orbit).

α-Particles are a highly ionizing form of particulate radiation that has low penetration. Each particle consists of two protons and two neutrons, which is identical to a helium nucleus; hence, the α-particle can be written as He^{2+}. α-Particles can be stopped by a thin sheet of paper or by the dead layer of the skin. Although not an external radiation hazard, α-particles released by radionuclides are dangerous if they are taken into the body by inhalation (breathing in) and/or ingestion (eating and drinking), as discussed further below. The adverse health effects caused by radon, an α-emitter, are explained by α-particles that are absorbed in the lung, thus becoming an internal radiation source. Indoor radon exposure can lead to lung cancer. Exposure from radon in drinking water is also of concern. Many countries have set exposure limit recommendations.

β-Particles are electrons that emanate from the nucleus of an atom. There are two forms of β-particles. They are the electron, given the symbol β^-, and the positron, given the symbol β^+. The depth to which β-particles can penetrate the body depends upon their energy. High-energy β-particles of several MeV may penetrate approximately 1 cm of tissue, although most are absorbed in the first few millimetres. As a result, β-emitters outside the body are hazardous only to surface tissue such as the skin of the lens of an eye. When β-emitters are taken into the body they irradiate internal tissues and become a more serious hazard.

The effect of radiation depends on the amount received and the exposure time. The amount of radiation received is expressed as a dose, and the measurement of dose is known as dosimetry. What is important is not so much the total dose to the whole system as the dose per mass of body tissue. This unit is the gray (Gy), equal to $J\,kg^{-1}$, and named in honour of the British physicist Louis Gray. The gray is a large dose, and for most normal situations we use the milligray (mGy) and the microgray (μGy). The absorbed dose is given the symbol D.

The gray is a numerical unit that quantifies the physical effect of the incident radiation (the amount of energy in joules deposited per kilogram), but it tells us nothing about the biological consequences of such energy deposition in tissue. Studies have shown that α- and neutron radiation cause greater biological damage for a given energy deposition per mass of tissue than, for example, γ-radiation. One Gy of α- or neutron radiation is more harmful than 1 Gy of γ-radiation.

Weighting factors are used to compare the biological effects of different types of radiation. For example, fast neutron radiation is more damaging than X-rays or γ-radiation. This leads to the idea that weighting factors can be used

Table 4 Quality factors for the various types of radiation.

Radiation	*Energy*	*Q*
gamma	all	1
beta	all	1
neutrons	slow	5
neutrons	fast	20
alpha	all	20

to quantify the fact that fast neutrons are more biologically damaging by expressing the statement that a lower absorbed dose produces equivalent biological effects. This is expressed with a radiation weighting factor, W_R, which is a function of energy deposition. The W_R of a certain type of radiation is related to the density of the ion tracks it leaves behind in the tissue; the closer the ion pairs, the higher the weighting factor. Another weighting factor used to further express biological effects is the tissue weighting factor, W_t.

Table 4 lists the W_R for various types of radiation. The values are valid for relatively long-term exposures; they do not apply to very large life-threatening doses received in a short time such as minutes or hours.

The absorbed radiation dose, when multiplied by the W_R of the radiation delivering the dose, will give us a measure of the biological effect of the dose. This is known as the dose equivalent, or dose equivalent index. Dose equivalent is given the symbol H. The unit of H is the sievert (Sv), also equal to $J\,kg^{-1}$. It was named after the Swedish scientist Rolf Sievert. An equivalent dose of 1 Sv represents that dose of radiation that is equivalent, for specified biological damage, to 1 Gy of X- or γ-rays. In practice, the millisievert (mSv) and microsievert (μSv) are the units in common use. Dose equivalent, quality factor and absorbed dose are related by:

$$H = DW_R$$

Most of the instruments used to measure radiation doses or dose rates display the values in mSv or μSv or in $mSv\,h^{-1}$ or $\mu Sv\,h^{-1}$, respectively. The collective dose to which a population is exposed is commonly quoted in 'man-sieverts' (man-Sv). However, this term should be avoided since it confuses a physical quantity with its units. Thus, sieverts alone is sufficient. The natural background effective dose rate varies considerably from place to place, but typically is around 3.5 $mSv\,year^{-1}$. For comparison, more than 6 Sv will lead to death in less than two months in more than 80% of cases, and much over 4 Sv is more likely than not to cause death.

For non-ionizing radiation, exposure standards are based on a measurement called the 'specific (standard) absorption rate' (SAR). The specific absorption rate is defined by the Institute of Electrical and Electronics Engineers (IEEE) as:

> the time derivative of the incremental energy (dW) absorbed by (dissipated in) an incremental mass (dm) contained in a volume element (dV) of a given density.

The specific absorption rate as defined by the American National Standards Institute (ANSI) reads:

> SAR is the time rate at which radio frequency electromagnetic energy is imparted to an element or mass of a biological body. SAR is expressed as energy flow (power) per unit of mass in units of W/kg.

When referring to human tissue, this means that the SAR is a measurement of the heat absorbed by the tissue.

Absorption in Toxicology. In toxicology, we are mainly concerned with absorption as given in definitions 3 and 4 above. In other words, we are concerned with the processes by which a chemical crosses the various membrane barriers of a living organism and especially those processes by which a chemical is taken up from environmental media, including food and other ingested material, such as drinking water, liquid refreshments and medication of all kinds. Absorption is the first step in the series of processes analysed in the study of toxicokinetics. The others are distribution (including storage), metabolism and excretion. Together, these are usually referred to by the acronym ADME (absorption, distribution, metabolism and excretion).

The main barrier to uptake is the phospholipid bilayer that forms the core of biological membranes. This prevents passive diffusion of water and water-soluble molecules but permits passive diffusion of fat-soluble molecules. Passive diffusion is driven by the electrochemical gradient of the substances involved. Fat-soluble substances move down their gradient of chemical activity (proportional to concentration of the substance) and/or the gradient of electrical charge (positive or negative).

Chemicals (foods, medicines, drugs of abuse, industrial chemicals and environmental chemicals) can enter the human body by various routes following ingestion, inhalation, injection (intravenous, subcutaneous, intramuscular), skin application, use of suppositories and uptake through mucous membranes of the eye or oral or nasal cavities.

Except for injection directly into the bloodstream, chemicals must pass through a complex system of cell membranes before they can enter the bloodstream. For example, chemicals that enter the digestive tract in solution or after solubilization may be absorbed by the cells lining the small intestine and then transferred through the cell to the other side (the transcellular route; Figure 18) where they cross the endothelial barrier into the bloodstream. Some chemicals may also pass between the epithelial cells of the intestine by what is called the paracellular route. Such transport is restricted by the junctions between cells, and this provides selectivity for the chemicals that can use this route. Chemicals that are inhaled must pass through the alveolar cells to the adjacent capillary cells and through them to the bloodstream.

As chemicals pass into and out of cells they must cross the cell membrane. The membrane defines the shape of the cell, controls the chemistry of the cell interior by regulating passage of substances into and out of the cell, and acts as

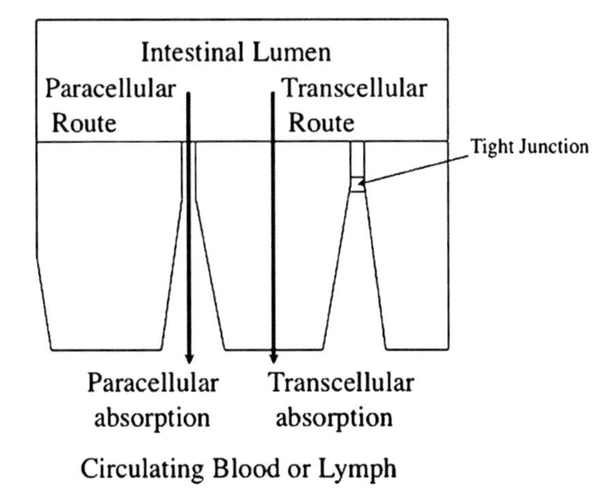

Figure 18 Transcellular and paracellular routes of absorption from the intestine.

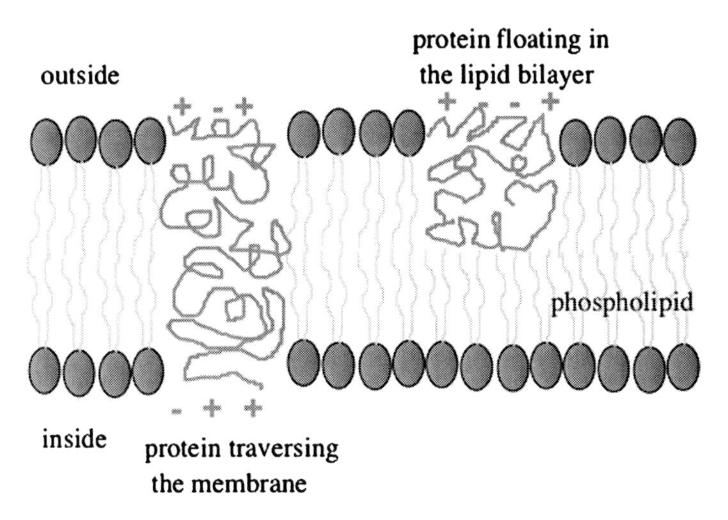

Figure 19 Fluid-mosaic cell membrane. Note the surface ionization, + or −, permitting pH-dependent ionic bonding at the outer and inner surfaces.

a transducer for extracellular chemical regulators such as hormones. The cell membrane consists mainly of phospholipid and protein as a lipid bilayer. Two phospholipid layers face each other inside the membrane with the more water-soluble parts of the phospholipid molecule (phosphate groups) facing the aqueous media inside the cell (cytoplasm) and outside the cell (extracellular fluid). The resultant structure is termed the fluid mosaic model (Figure 19), and the fluidity is crucial to its function. The membrane proteins provide some rigidity to the structure, and some may act as transporters. Some 'float' in the membrane, binding to external substances and diffusing with them from one side of the membrane to the other. Others may traverse the membrane structurally and, on binding to a substance, change shape so that the substance is transported across the membrane. This process may be associated with input of energy from nucleoside triphosphate breakdown, permitting transfer against an electrochemical gradient, referred to as active transport (see below).

The most fundamental mechanism for transport of either foreign chemicals or ordinary ions through the cell membrane is passive diffusion. The driving force for passive diffusion of a chemical is based on the difference between the concentration (or better chemical activity) of the chemical (or chemicals of the same sign of electrical charge if ionized and experiencing a charge gradient established by active transport, see below) on the outside of the cell and that on the inside of the cell. This is properly called the electrochemical gradient. The greater the difference in the relevant electrochemical activity between the outside and the inside of the cell, the greater the diffusion of the chemical down the resultant gradient, in or out. Since the membrane barrier to chemical movement is mainly lipid, the ability of a chemical to diffuse across the membrane is largely dependent on its lipophilicity (solubility in lipid). This is measured in practice as its octanol–water partition coefficient (K_{ow}, P_{ow}), *i.e.* the ratio of the concentration in octanol to that in water after the substance is mixed thoroughly with both and they are allowed to come to equilibrium. Often this is given as the log K_{ow}. It is generally considered that values above 3 indicate increasing tendency to absorption. At very high values of log K_{ow}, 6 or above, absorption may be restricted to the membrane phospholipid bilayer as such substances are too lipid soluble to pass further into inner aqueous media, either the cytoplasm or further aqueous fluids such as blood. Notably, log K_{ow} is not an accurate determinant of lipophilicity for ionizable compounds because it only correctly describes the partition coefficient of neutral (uncharged) molecules. Since most drugs (approximately 80%) are ionizable, log P is not an appropriate predictor of a drug's behaviour in the changing pH environments of the body. The distribution coefficient (log D) is the correct descriptor for ionizable systems. The distribution coefficient is the ratio of the sum of the concentrations of all forms of the compound (ionized plus unionized) in each of two phases in contact. For measurements of distribution coefficient, the pH of the aqueous phase is buffered to a specific value such that the pH is not significantly perturbed by the introduction of the compound. The logarithm of the ratio of the sum of concentrations of the solute's various forms in one solvent to the sum of the concentrations of its forms in the other solvent is called log D. Log D is pH dependent and, hence, one must specify the pH at which the log D was measured. Of particular interest is the log D at pH 7.4 (the physiological pH of blood serum). For un-ionizable compounds, log P=log D at any pH for which the compound remains unionized.

Because they are not lipid-soluble, charged molecules do not readily diffuse across the plasma membrane. The pH of the fluid surrounding the cell is important in this respect because it influences ionization and hence molecular charge. Weak acids in their un-ionized form may be lipid-soluble and will diffuse across membranes quite easily. The degree of ionization of a molecule at different pHs is dependent on its pK, *i.e.* the pH at which 50% of the chemical is ionized and 50% is un-ionized. This is important in the gut. In the human, and generally in the carnivore stomach lumen, the pH may be as low as 1.5 to 2.0. In other species (*e.g.* in rats), the pH may be much higher (*e.g.* pH 4). However, the pH inside the small intestine is about 7.0 to 8.0. Thus, the ratio of ionized to un-ionized chemical differs for any chemical in these two environments, depending on its pK, and the amount of ionizable chemicals absorbed from the stomach and the small intestine is different. This property may be used in designing drug molecules to ensure preferential absorption from the stomach by giving the molecule a structure with the appropriate pK.

Water-soluble biomolecules and other chemicals may be transported across the membrane with the aid of carrier proteins to which they become bound (see also carrier). One possibility is facilitated diffusion in which the molecules move down the electrochemical gradient exactly as happens with normal diffusion, although the process may be limited by the availability of carrier molecules and the kinetics of the binding–unbinding reaction between carrier and chemical. It is also possible for chemicals to compete for the binding sites on carrier proteins. A particularly important carrier is a multidrug resistance carrier, which has been demonstrated to transport a wide range of xenobiotics conjugated to glutathione, glucuronate or sulfate as well as unmodified anionic compounds such as the antifolate agent methotrexate.

Chemicals may also cross the cell membrane by diffusion through water-filled membrane pores. This diffusion is dependent on the size of the pore and the molecular size and shape of the chemical.

Some inorganic ions, such as sodium and potassium, and many drugs move through the cell membrane by a process called active transport. This process moves substances against the electrochemical gradient and requires input of energy, usually as adenosine triphosphate (ATP). Thus, active transport absorption by any cell will reflect its metabolic activity and, in some circumstances, may stress this so much as to have adverse effects. Figure 20 illustrates a good example of active transport, the sodium pump. ATPase binds Na^+ and ATP in the E_1 conformational state (step 1) and is phosphorylated at an aspartate residue by the γ-phosphate of ATP. This leads to the occlusion of three Na^+ ions (step 2) and then to their release to the extracellular side (step 3). This new conformational state (E_2-P) binds K^+ with high affinity (step 4). Binding of K^+ leads to dephosphorylation of the enzyme and to the occlusion of two K^+ cations (step 5). K^+ is then released to the cytosol after ATP binds to the enzyme with low affinity (step 6). The dashed box highlights the electrogenic steps of the catalytic cycle.

Finally, the membrane can engulf the chemical, form a vesicle and transport it across the membrane to the inside of the cell. This process is called endocytosis (Figure 21) and is especially important for particulates. The same process can occur in reverse and then is known as exocytosis. It is energy-dependent and may result in the transport of a mixture of chemicals because, although it is usually induced by a specific process, it may take unspecifically

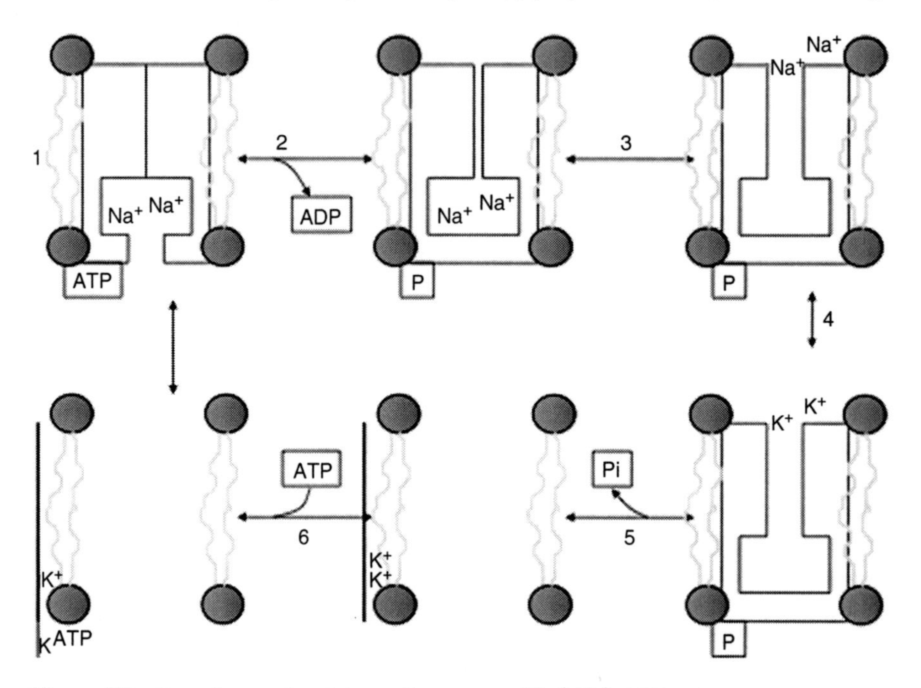

Figure 20 Reaction cycle of the sodium pump (Na^+/K^+-ATPase).

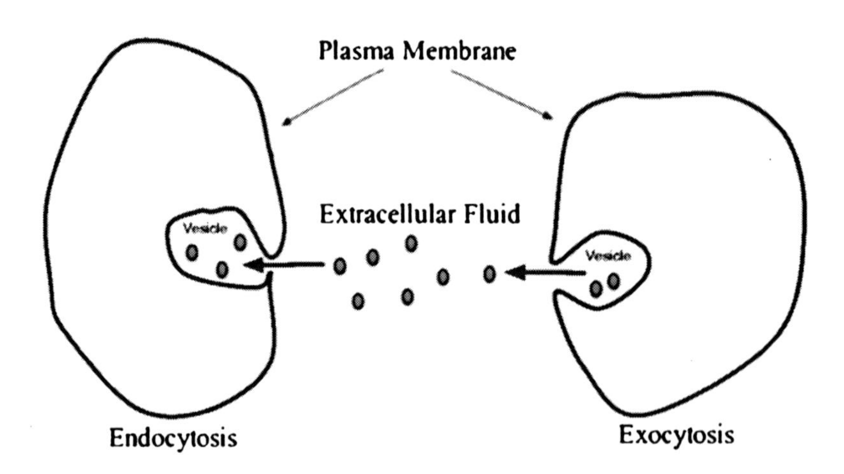

Figure 21 Endocytosis and exocytosis.

enclosed substances into the cell. This process has been suggested to be more active in the newborn.

Once the chemical has entered the bloodstream or the lymph it is distributed to organs distant from the site of absorption. Initially, simply because of the electrochemical gradients between the blood or lymph and the organs that it perfuses, many chemicals will tend to leave the blood (or lymph) passively and enter the surrounding cells (uptake). The pH of the blood (or lymph), pH 7.4, will determine the ionization state of polar organic chemicals, and this will influence passage through the cell membranes of the cells separating the blood (or lymph) from the organs.

Anatomical and physiological factors may affect the movement of a chemical around the body. For example, the cells surrounding the capillaries in the brain have tight junctions that impede the flow of materials between cells. One type of glial cell in the central nervous system, the astrocyte, forms a tight covering on the brain's capillaries and prevents or retards large molecules from entering the brain. This structure constitutes what is known as the 'blood–brain barrier'.

The placenta is an organ that permits nutrients to pass from the mother's bloodstream to that of her foetus but does not allow the passage of all chemicals. The maternal blood and the foetal blood do not have direct contact. Generally, molecules with a molecular mass greater than 1000 Da have difficulty entering the foetal blood supply. The placenta can metabolize chemicals, and the derivatives produced may be responsible for effects on the embryo and foetus.

Other factors to be considered regarding the availability of chemicals to cells in the human body are chemical affinity and the resultant long-term incorporation into tissues such as fatty tissue and bone. Chemicals that are lipophilic have an affinity with, and a tendency to be absorbed by, and accumulate in fat cells, from which they are released very slowly under normal circumstances. Such chemicals may cause no problems for years, but the fat cells can break down quite rapidly during pregnancy or illness or in old age. Any stored lipophilic xenobiotics may then flood the bloodstream and cause illness as a result of uptake of toxic doses into susceptible organs, particularly into the critical organ. The same effect occurs in birds and wild animal species during the winter when food supplies are short.

Affinity with proteins in the blood may also be important in determining the availability of chemicals to susceptible tissues. In particular, some chemicals may become strongly bound to plasma proteins (such as albumin), and the rate of release from such binding will determine how long the chemical is available to exert its biochemical and physiological effects. Serum albumin is important in the transport of bile pigments such as bilirubin. Some drugs, such as sulfasoxazole and ceftriaxone, can compete for bilirubin-binding sites on the albumin molecule, causing it to be released and deposited in tissues where it causes damage. This is just one example of many hundreds that are of clinical importance.

For some inorganic species, such as fluoride ion, lead ion and strontium ion, incorporation into bone may occur. The elements may stay there for long periods of time. As bone slowly renews itself or is partly broken down during pregnancy, illness or old age, these chemicals may be released. If this occurs during pregnancy, the resultant toxicity can affect both the mother and her child. Similar toxic effects may be seen in the sick and the elderly. Since these effects may occur some time after exposure to the chemicals, diagnosis of the cause depends upon their detection in blood or urine samples.

For some toxic substances, it is important to evaluate if the adverse effect (*i.e.* the toxicity caused by the substance) results from its radioactivity or from its direct toxicity as a result of its other properties, or from both. For example, ^{203}Hg and ^{109}Cd may cause damage as a result of both radiation effects and toxic effects due to direct chemical reactions.

Further Reading

National Library of Medicine (Content source); Emily Monosson (topic editor), 2008, 'Absorption of toxicants', in *Encyclopedia of Earth*, ed. C. J. Cleveland, Environmental Information Coalition, National Council for Science and the Environment, Washington, D.C. (First published in the *Encyclopedia of Earth* October 6, 2006; last revised February 28, 2008; retrieved August 21, 2008). Available at <http://www.eoearth.org/article/Absorption_of_toxicants>

3.3 Compartment

compartment
Conceptualized part of the body (organs, tissues, cells, or fluids) considered as an independent system for purposes of modelling and assessment of distribution and clearance of a substance.
compartmental analysis
Mathematical process leading to a model of transport of a substance in terms of compartments and rate constants, usually taking the form $C = Ae^{-\beta t} \cdots$, where each exponential term represents one compartment. C is the substance concentration; A, B, . . . are proportionality constants; α, β, . . . are rate constants; and t is time.

A compartment as used in toxico- or pharmacokinetics is an abstract concept that should be thought of in purely arithmetical terms; although the mapping of this concept onto a biological structure may be straightforward and in some cases may seem obvious, this is by no means the way to understand compartmental analyses, as the compartments need not have any anatomical reality. The need for compartments, then, is guided by the need to analyse the movement of substances, their accumulation and uptake, their clearance and elimination, and their redistribution. Because we often wish to describe a concentration of a substance and its change (*e.g.* the derivative of substance amount with respect to time) we need a representation of volume in the denominator. Thus, a compartment is, in its broadest meaning, any volume that can be used to define the concentration of a substance of interest.

General Considerations. Typically, one of the most useful compartments is the circulating blood volume, which may be defined specifically as blood (the fluid including all cellular components), or as the plasma (free of cells) or serum (plasma free of coagulating proteins). In this context, the blood behaves as a compartment, and we can consider the entry of a substance into this compartment or space, or its elimination therefrom. From the principle that first-order chemical reactions will occur at rates proportional to the amount of the reactant, a simple exponential equation usually describes the rate of appearance or disappearance in the compartment. A simple example of compartmental behaviour, then, is the exponential rate of decay of a substance in the blood, *i.e.* its disappearance from the blood.

If compartments can be considered as hypothetical spaces between which substances move according to defined kinetic principles, we can introduce the idea of compartmental analysis as the mathematics that describes this movement. Complex multi-compartment models can be employed to model complicated behaviour, but the general principles are well illustrated by considering the cases of one- and two-compartment systems. Surprisingly, the intricate biology that determines the full toxicokinetics of a substance is often modelled very well by these simple compartmental descriptions based on first-order kinetics.

The simplest one-compartment model depicts the whole body as a single homogeneous unit. If a substance distributes rapidly throughout the body, this might be a rather accurate model, and the plasma might be the compartment of choice to describe the substance's behaviour, because of its accessibility for sampling. Note that the criterion of homogeneity does not mean that the substance is assumed to be uniformly distributed throughout the body; rather, kinetic homogeneity means that the rate of the substance's appearance in, or disappearance from, the plasma well represents its rate of appearance or disappearance in any other part of the body. Important concepts are the maximum concentration reached in the compartment (C_{max}) and the time at which it occurs (t_{max}). The integration of the curve of concentration of the substance versus time, from zero to infinity (area under the curve), depends upon the total amount of the substance that has entered the compartment, and is useful for describing dose or exposure. The rate, K, of addition to, or elimination from, a compartment is the sum of individual rate constants (k_i) of the various processes. In the example of an infused drug, X, of concentration [X], it is common to write the elimination rate as $d[X]/dt = k - K[X]$, where k is the rate of infusion. Often in pharmacokinetics, one uses the Laplace transform to solve this and similar differential equations to eliminate the time variable, yielding equations that can be solved more easily. Two-compartment (and higher-order) models are useful for describing exchange between pools. In the simplest sense, a pair of exchange rate constants, k_{12}/k_{21}, describe the flux between compartments, and an elimination rate constant, k_{10}, accounts for removal of the substance from a single compartment. As additional compartments and multiple sites of uptake and/or elimination come into play, the arithmetic becomes more complicated, but the underlying principles, and their utility, remain the same. The description of transfer of a substance among compartments facilitates modelling of kinetic behaviour, so-called pharmacokinetic (PK) modelling. The resulting models are descriptive, giving a concise mathematical description of empirical data, for instance. Predictive models, on the other hand, accommodate nonlinear aspects of ADME (Section 3.2) that are not well handled by classical compartment models. PBPK (physiologically-based pharmacokinetic modelling) is one of the most successful among predictive models. Proposed mechanistic aspects of physiological phenomena are incorporated, and rigid assumptions, *e.g.* of steady-state or first-order kinetics, are avoided. For instance, successful PBPK (PBTK, physiologically-based toxicokinetic modelling) models of lead, chromium and methylmercury in humans has been developed that takes into account perfusion rates of various tissues where redox reactions, lipophilic partitioning or selective ligation can occur.

Note: The fundamental compartment concept can be applied to mathematical modelling of distribution of chemicals in the environment and this has been done most successfully by Mackay using fugacity to assess movement from one environmental medium to another (Section 4.1).

Further Reading

IPCS, *Human Exposure Assessment,* Environmental Health Criteria 214, WHO, Geneva, 2000. Available at International Programme on Chemical Safety (IPCS) INCHEM (Chemical Safety Information from Intergovernmental Organizations), 2008. Available at <http://www.inchem.org/>.

IPCS, *Principles of Toxicokinetic Studies*, Environmental Health Criteria 57, WHO, Geneva (1986). Available at International Programme on Chemical Safety (IPCS) INCHEM (Chemical Safety Information from Intergovernmental Organizations), 2008. Available at <http://www.inchem.org/>.

M. Makoid, P. Vuchetich and U. Banakar, *Basic Pharmacokinetics*, 2008. Available at <http://pharmacyonline.creighton.edu/pha443/pdf/default.htm>.

D. D. Shen, Toxicokinetics, in *Casarett & Doull's Toxicology – The Basic Science of Poisons*, ed. C. D. Klaasen, McGraw-Hill, New York, 7th edn, 2008, p. 305.

3.4 Volume of Distribution

volume of distribution

Apparent (hypothetical) volume of fluid required to contain the total amount of a substance in the body at the same concentration as that present in the plasma, assuming equilibrium has been attained.

Volume of distribution can be expressed as:

V_d/litre = (dose mg^{-1})/(plasma concentration/mg L^{-1})
or=(dose mol^{-1})/(plasma concentration/mol L^{-1})

From this relationship, lower plasma concentrations imply a higher volume of distribution of the substance while higher plasma concentrations imply a lower V_d. A V_d for a given substance of about 5 L would imply the substance is primarily in the plasma. In contrast, a V_d of much more than 5 L implies that the substance is more widely distributed through the body. A value of more than 50 L indicates that the compound is accumulated in the body.

General Considerations. If a toxic substance is mostly bound to plasma proteins such as albumin, the V_d will approximate to the plasma volume. If a toxic substance is highly lipid soluble, and distributes mainly to adipose tissue, the plasma concentration will be low and the V_d will be larger than the plasma volume and may even exceed the volume of total body water.

The V_d has certain limitations. The volume of distribution is a theoretical measurement, and the possibility that it may exceed the volume of total body water emphasizes this fact. Toxic substances have different affinities for different body tissues, and the observation of a large V_d does not indicate the location of the relevant toxic substance in the body. Even where this is known, it must be remembered that the main location of the substance may not be its site of action. For example, organochlorines accumulate in fatty tissue, but their site of action may be on the nervous system or on the reproductive system.

Plasma concentration, and hence volume of distribution, changes over time, and so a single determination of V_d gives much less information than a time course study.

Further Reading

D. D. Shen, Toxicokinetics, in *Casarett & Doull's Toxicology – The Basic Science of Poisons*, ed. C. D. Klaasen, McGraw-Hill, New York, 7th edn, 2008, p. 305.

3.5 Half-life

half-life, $t_{1/2}$ (half-time, $t_{1/2}$)
Time required for the concentration of a reactant in a given reaction to reach a value that is the arithmetic mean of its initial and final (equilibrium) values. For a reactant that is entirely consumed, it is the time taken for the reactant concentration to fall to one-half its initial value.

Note: The half-life of a reaction has meaning only in special cases:

1. For a first-order reaction, the half-life of the reactant may be called the half-life of the reaction.

2. For a reaction involving more than one reactant, with the concentrations of the reactants in stoichiometric ratios, the half-life of each reactant is the same, and may be called the half-life of the reaction.

If the concentrations of reactants are not in their stoichiometric ratios, there are different half-lives for different reactants, and one cannot speak of the half-life of the reaction.

The concept of half-life originated in relation to the decay of radioactivity. Subsequently, it was applied in biology to define the rate of removal of substances from the body and, most recently, it has been used to define rate of disappearance of substances from environmental media as described below.

Radioactive Half-life. Half-life in toxicology can be either radioactive or biological. Radioactive half-life is the time taken for half the number of atoms in a radioactive substance to decay. Different radionuclides have different half-lives and emit different forms of radiation and thus have different toxicological effects. In radiation toxicology both types of half-life must be taken into account. If a radionuclide has a long radioactive half-life but a short biological half-life in the tissues or cells of a given organism, the chances of harm resulting from the radioactivity in that organism are small because the radionuclide will probably be excreted before significant radioactive decay occurs. But the possibility of radiation damage is still there, however small it may be. An even more complex situation arises if the decay process follows a cascade mechanism through intermediate radioactive daughter isotopes. In this situation, every case must be considered individually and no generalizations are possible.

Biological Half-life. At its simplest, the biological half-life is the time required for the amount of a particular substance in an organism to be reduced to one-half of its value by biological processes when the rate of removal is approximately exponential. Substances with a long biological half-life will tend to accumulate in the body and are, therefore, particularly to be avoided. Knowing that they accumulate in the body is useful but not in itself sufficient to assess the consequences and to take steps to minimize them. Different organs have different half-lives for the presence of the same substance. For example, methylmercury has been reported to have a half-life of about 50 days in the liver and 150 days in the brain. Thus, the tendency for accumulation in the brain is greater and so is the risk of brain damage. Since the overall half-life for methylmercury in the human body is only 70 days, it is clear that there is a danger of underestimating risk if only total body half-life is assessed.

An increasingly important aspect of half-life determination is the half-life for disappearance from the natural environment, especially of man-made toxicants. Persistence in the natural environment is seen as a substance property that must eventually lead to problems simply because all substances are toxic at a given concentration. The toxic concentration will eventually be reached for even the least toxic substance if it persists sufficiently long and is constantly being added to the environment. This concept underlies the definition of persistent organic pollutants (POPs), which are defined as persistent if:

1. there is the potential for long-range transboundary atmospheric transport (necessary evidence includes a half-life in air of > 2 days);
2. there is aquatic persistence: half-life in water > 2 months;
3. there is persistence in soils: half-life in soil > 6 months;
4. there is persistence in sediments: half-life in sediments > 6 months.

There is a problem in applying these criteria in that laboratory tests to establish half-lives for organic pollutants cannot simulate all the environmental possibilities contributing either to breakdown or stabilization of compounds. In particular, since most organic breakdown in the natural environment is due to microorganisms, the presence or absence of appropriate microorganisms is often the determining factor in half-life assessment and so there may be a large difference between laboratory estimates and environmental behaviour simply because of variability in this factor.

Further problems arise in assessing half-lives of inorganic compounds in the natural environment. While 'mineralization', the breakdown of organic compounds to carbon dioxide and water, is considered the 'harmless' end of their natural life, no similar criterion is available for inorganics. Historically, most inorganics containing metallic elements have been defined toxicologically in relation to their content of the metallic element. Thus, toxicologists have talked loosely of the toxicity of chromium or nickel when in fact most of the toxicological effects relate to specific forms of these metals such as chromate anions or nickel tetracarbonyl. In consequence, regulatory authorities have regulated for levels of the element in environmental media, sometimes defining a relevant oxidation state, as with chromium(VI). Thus, since elements are persistent by definition, all inorganic elements regulated in this way automatically become persistent inorganic pollutants (PIPs). However, where testing has been done, most simple elements in the pure elemental state show low toxicity, mainly because they are not bioavailable. Thus, there is a need for a better approach to regulation of inorganic pollutants incorporating an appropriate method for determining environmental half-life of those compounds that are genuinely toxic.

Substances with a short biological half-life may nevertheless accumulate if a small amount becomes tightly bound or the organism makes more receptor molecules, even if most is cleared from the body rapidly. There is also the possibility that substances with a short biological half-life may have cumulative toxic effects. This is the most difficult situation for the toxicologist to interpret, but it may be quite common in long-lived organisms such as human beings. Thus, care must be taken to understand the toxicokinetic metabolism of any substance before regarding it as relatively harmless on the basis of a short half-life.

Similar reservations must be applied to consideration of substances with short half-lives in environmental media. Half-life is measured in relation to the bulk of the chemical present, and it may be that a small fraction persists in association with a component of the environment that is small in quantity but ecologically important or essential for a species that has colonized a unique niche. Again, interpretation of half-life for risk assessment must be cautious and based on as full as possible an understanding of environmental chemistry and ecology.

Further Reading

D. J. Greenblatt, Elimination half-life of drugs: value and limitations, *Annu. Rev. Med.*, 1985, **36**, 421.

J. G. Wright and A. V. Boddy, All half lives are wrong, but some are useful, *Clin. Pharmacokinet.*, 2001, **40**, 237.

3.6 Biological Monitoring (Biomonitoring)

biological monitoring
Biological assessment of exposure.
biomonitoring
Continuous or repeated measurement of any naturally occurring or synthetic chemical, including potentially toxic substances or their metabolites or biochemical effects in tissues, secreta, excreta, expired air or any combination of these in order to evaluate occupational or environmental exposure and health risk by comparison with appropriate reference values based on knowledge of the probable relationship between ambient exposure and resultant adverse health effects.

Because of the uncertainties involved in estimating environmental exposures, *i.e.* external exposure through an environmental medium and the resultant internal dose, there is an increasing trend towards assessing exposures of living organisms to chemicals in their environment by biomonitoring. This essentially refers to the analytical process of measuring the concentration of substances or their metabolites in blood, urine, breast milk, hair and other biological samples taken from exposed organisms. Biomonitoring should, in theory, be a better measure of exposure than analysis of environmental media since it relates to the internal dose produced by external exposure of the organism being studied. However, this relationship is affected by many environmental and physiological factors. Interpreting biomonitoring data is, therefore, often difficult unless there is clear evidence that such factors can be ignored.

Conventional environmental exposure scenarios often use 'worst case' assumptions. They have been designed to provide estimates of maximum possible external exposure that can be allowed while still protecting the organisms at risk from harm. Thus, such assessments are likely to overestimate actual exposures. Biomonitoring provides values that are a direct measure of the dose resulting from an individual's integrated exposures from multiple pathways and sources. However, biomonitoring cannot identify specific sources or pathways of exposure or the relative contributions from multiple sources.

Objectives of Biomonitoring. Biomonitoring studies are used for the following purposes:

1. to determine which chemicals are taken up by living organisms and at what concentrations;
2. to determine the prevalence of organisms with exposures likely to cause toxicity;
3. to establish reference concentrations;
4. to assess the effectiveness of attempts to reduce exposure;
5. to compare exposure levels in different groups;
6. to track trends in exposure over time;
7. to set priorities for research.

Achieving these objectives is facilitated by the maintenance of specimen banks under appropriate conditions to prevent deterioration of samples. Specimens must be well documented and defined precisely in relation to their origin and to all factors that might affect interpretation of relevant data. This will permit re-investigation of specimens as analytical techniques improve and new substances of concern are identified.

Biomonitoring data may be used to determine whether individuals or a population are at an increased risk of experiencing adverse health effects associated with an exposure to a specific substance. Criteria for the evaluation of biomonitoring data in this context have mostly been developed for exposures in the workplace. In the United States, the American Conference of Governmental Industrial Hygienists (ACGIH) began developing biomonitoring-based reference values known as biological exposure indices (BEIs) in the early 1980s (*see* Further Reading). The ACGIH defines the BEIs as representing the levels of determinants that are most likely to be observed in specimens collected from a healthy worker exposed to chemicals following inhalation exposure at the threshold limit value (TLV). The Deutsche Forschungsgemeinschaft (DFG) in Germany has also developed biological monitoring reference values called biological tolerance values (BATs), and the World Health Organization has established similar values that they call biomonitoring action levels (BALs). BEIs, BATs and BALs all refer specifically to occupationally exposed populations and exposures in occupational scenarios only.

For the general population, health-based screening levels for biomonitoring data exist for very few substances (exceptions being lead, mercury, arsenic, cadmium and ethanol). At present, almost all regulatory health-based toxicity screening criteria are based on an estimated intake level (mg/kg body mass/day) or a concentration in an environmental medium (air, water, soil, *etc.*) that corresponds to what is regarded by an expert group as an acceptable level of intake.

Analytical Aspects of Biomonitoring. The first paper on biomonitoring was published in 1927 (*see* Further Reading), and presented the analysis of lead in the urine of exposed workers as a means of diagnosing lead-induced occupational disease. A well-designed sampling program and accurate chemical analysis are essential for correct interpretation of the analytical results obtained. Although analytical techniques have developed with time, analytical quality is a major concern for all biological monitoring programs, and trustworthy external quality control must be established before any program starts. Analyses should be performed by an accredited laboratory.

Biomonitoring of occupational exposure to chemical substances differs from normal clinical chemistry analyses in the dependence of the measured concentration of the chemical on toxicokinetics and exposure patterns. At work, the exposure is generally assumed to occur for a workshift period of 8 hours daily and may be limited to only short periods of time within the working hours. These assumptions vary with different regulatory authorities and must be known for biomonitoring data to be interpreted properly. Some chemicals or their metabolites have a very short half-life in the body, especially in blood, and thus the concentration drops very rapidly immediately after the exposure. Thus, the concentration measured may reflect the time lapse between exposure and sample collection more than it reflects the original maximum concentration following exposure. This may lead to a distorted assessment of the exposure – and thus of the risk involved. The conclusion could even be that there was no exposure although exposure may have occurred, but, by the time the sample was collected, all of the substance of concern had disappeared from the blood.

Use of Biomarkers in Biomonitoring. Biomarkers are discussed separately below (Section 3.8). Biomarkers of exposure may be used to identify exposed individuals or groups. This depends on comparing the results with established reference levels, or with biomonitoring action limits, if these have been defined. A biomarker level that occurs at a concentration above a reference level indicates that the individual monitored has been exposed to a greater extent than the reference population but is not of itself a measure of ill-health. However, it is a warning of a possible health hazard that requires further investigation. Biomarkers of exposure take no account of inter- or intra-individual differences in the toxicodynamics of the chemical. Such differences may be identified by biomarkers of effect, bearing in mind that specificity of the relationship of the effect to the exposure of concern must be clearly established.

Biomarkers do not differentiate between sources of exposure, and in order to decrease the risk from the chemical that is monitored it may be necessary to consider (and to analyse) separately whether the exposure occurred at work or at home. Biomarkers of exposure in humans usually reflect the amount of the chemical in the systemic circulation, and models have been developed to predict concentrations in other compartments in the body from such data. However, a major difficulty in the interpretation of biomonitoring data from humans relates to the concentration and effects of substances at the site of entry, *e.g.* effects on the lungs after exposure to particulates containing metallic elements such as nickel. Concentrations of nickel ions in the urine or the blood may reflect the concentrations in, or the health risks to, the lungs after exposure to such particles, but only inadequately since the nickel compounds most associated with lung cancers are largely insoluble. Nor do they indicate the exact chemical speciation of nickel in the lungs, which must be known to assess the risk of cancer developing since the risk varies hugely with the nickel species, with nickel salts posing little if any risk, dependant, of course, on exposure concentration.

Biomarkers of effect have the intrinsic advantages over biomarkers of exposure in that they reflect differences in individual sensitivity to the chemical. Thus, for example, in exposure to cadmium, assessment of excretion of low molecular mass proteins in urine may be used to identify individuals who are exceptionally sensitive, and who develop adverse health effects at levels of exposure at which individuals with normal sensitivity remain healthy. Cadmium-related renal toxicity is a well-established example where such an advantage can be achieved. For such toxicity, an increasing number of proteins and enzymes and expression of mRNA for a few proteins, *e.g.* metallothionein, have been identified to serve as biomarkers of effect and also to identify vulnerable groups in the population.

Regarding biomarkers of effect, the availability of highly sensitive analytical methods has provided insight into the formation, persistence and repair of DNA adducts induced by numerous chemicals. By understanding the molecular dose of such adducts in different cells and tissues we have learned the metabolism and mode of action of a range of substances across species. In addition to DNA adducts, similar information has been obtained from studies of protein adducts and related chemical metabolites in urine and plasma. For example, much research has been conducted on the molecular dosimetry of aflatoxin in rats and humans, with measurements of DNA adducts, protein adducts and urinary excretion of both adducts and metabolites. The earlier literature has been reviewed by Busby and Wogan (*see* Further Reading). Aflatoxin B1 (AFB1) forms adducts at the N-7 position of guanine. These adducts can depurinate and be excreted in the urine, but they also form the ring-opened FAPY adduct, which is persistent and mutagenic and therefore carcinogenic.

Chemical Speciation in Biomonitoring. If we ignore carbon and its derivatives, routine biomonitoring of all other elements has in the past been almost entirely dependent on the analysis of the total content of the element as a biomarker, without any consideration of the different chemical species in which the element may be present. For some elements, this straightforward approach may be sufficient. It is probably applicable where the key effect of the element on health is caused by its most common ionic form in aqueous media and the dose–response relationship between the total element concentration and any beneficial or adverse health effect is known. Thus, a reliable prediction of long-term health effects may generally be made from total lead in blood or total cadmium in blood or urine. However, if we are dealing with exposure to tetraethyl-lead in air or to methylmercury in the diet, total elemental analysis will be seriously misleading for any health risk assessment. Any error will be compounded if there is concomitant exposure to different chemical species of the same element by different routes and under different circumstances, *e.g.* exposure to inorganic arsenic species in the workplace and organic forms in the diet.

Toxicity is dependent on chemical speciation as much as on the organism at risk. Thus, chemical speciation analysis is essential for biomonitoring, whether of human beings or any other biological species. Chemical speciation analysis requires the development of three different approaches:

1. fractionation analysis (*e.g.* to separate organic and inorganic forms of arsenic);
2. speciation analysis, to precisely define individual chemical species;
3. analysis of the distribution of different species in tissues and organs (*e.g.* mercury in plasma, blood cells and urine; chromate in erythrocytes and plasma).

If only the total concentration of an element is measured, exposure indices used for comparisons to assess health risks should relate this to defined chemical species to which the population has been exposed, and which have been characterized for their dose–response relationships to the health effect or effects of concern. At present such exposure indices are difficult to find and they will need to be developed for most common elements.

Biomonitoring Equivalents. Most existing chemical risk assessment procedures rely on external exposure measurements to set exposure guidance values such as reference doses or tolerable daily intakes. However, based on metabolism and kinetics of the substance concerned, attempts have been made to establish what have been called biological monitoring guidance values referring to internal concentrations in humans, deduced for blood or urine samples. For example, human health guidance values (HGVs) have been defined by the U.K. Health and Safety Executive (HSE) as levels of a substance or its metabolites in blood, or urine, that are not associated with any adverse health effects. Another type of biological monitoring guidance value defined by HSE is called a 'benchmark value'. This type of value is set when it would not be appropriate to set a HGV – for example, for substances that can cause cancer. The benchmark value is based on a survey of workplaces that are considered to have good control of exposure to the substance and it is the value found in 9 out of 10 samples in those workplaces. This type of guidance value can give no direct guide to the risk of ill-health. The benchmark guidance value just gives an indication of how well exposure is being controlled, and should be used as a trigger for further investigation. A similar approach has been adopted by the American Conference of Governmental Industrial Hygienists (ACGIH) (*see* Further Reading), which has established a series of recommended reference values for biomonitoring called the biological exposure indices (BEIs). BEIs are defined as reference values intended as guidelines for the evaluation of potential health hazards in the practice of industrial hygiene. BEIs represent the levels of determinants that are most likely to be observed in specimens collected from a healthy worker who has been exposed to chemicals to the same extent as a worker with inhalation exposure to the TLV.

At present, few reference values are available to interpret the increasing amount of available human biomonitoring data. To solve this problem, the basic concept of biomonitoring equivalents (BEs) for the general population, *i.e.* biomonitored blood or urine concentrations of chemicals corresponding to existing exposure guidance values, was introduced by Hays *et al.* (*see* Further Reading), integrating toxicokinetic data with existing chemical risk assessments to provide such reference values.

Some of the methods that can be used to develop BEs and reviewed by Hays *et al.* are:

1. extrapolation from occupationally derived biomarker levels such as the biological exposure indices (BEIs) set by the ACGIH;
2. human PK studies and one-compartment steady-state models;
3. multi-compartment and PBPK models;
4. animal PK studies.

In general, the methods focus on blood as the medium of interest, although similar approaches can be used to convert into BEs for use with other sampled media. Interpretation of data for each medium requires consideration of issues specific to that medium. For example, urine is frequently sampled for metabolites from exposures to a wide range of drugs and chemicals, and appropriate methods for standardization of urinary output volumes must be considered.

A Biomonitoring Equivalents Expert Workshop was held in 2007 and this resulted in the production of detailed guidelines for the derivation of Biomonitoring Equivalents that were published in 2008 (*see* Further Reading).

Ethical Questions Related to Biomonitoring. The following questions should be considered in relation to any proposed biomonitoring project:

1. Is biological monitoring justified by the predicted outcome(s)?
2. Can the monitoring procedure cause any harm to participants?
3. Have participants been fully and appropriately informed about the aim and methods of the biomonitoring program and the proposed use of any information gathered during the study?
4. If the data indicate potential health problems for the participants in future, how should they be informed?
5. Have the participants given informed assent or consent and been given the right to withdraw subsequently if they have any doubts?

6. Are there possible health implications for any group to which the participant belongs?
7. Are there systems in place for communication of results to participants?
8. Will participants have access to their own data and have they been informed about relevant data protection including who has the right to know or not to know?
9. If the data have commercial value, who will benefit from this?

Conclusions. Further development of biomonitoring will depend upon the following factors:

1. further development of inexpensive monitoring methods that make possible adequate biomonitoring of substances with a short half-life in the body;
2. derivation of exposure biomarker guidance values for defined elemental species that are serious risk factors;
3. identification of fully validated biomarkers of effect;
4. validation and application to effect-monitoring of the -omic technologies (Section 2.11);
5. validation of the appropriateness of media and markers to be measured in proposed biomonitoring studies;
6. insistence on quality control at all stages of biomonitoring, especially relating to possible contamination of samples during collection;
7. improved definition of the stability (or lack thereof) of within-subject biomarkers to relevant exposures over time (especially in relation to the often unstated assumption that one measurement adequately reflects likely exposure during a critical period);
8. improved definition of between-subject variability in biomarkers and body burden after apparently similar exposures;
9. improved understanding of differences in the toxicokinetics of specific subgroups (*e.g.* relating to age and sex);
10. continuing development of the knowledge base relating to species differences in toxicokinetics and toxicodynamics.

Further Reading

A. Aitio, *Clin. Chem.*, 1994, **40**, 1385. <http://www.clinchem.org/cgi/reprint/40/7/1385.pdf>.

American Conference of Governmental Industrial Hygienists (ACGIH) (2008). Available at <http://www.acgih.org/home.htm>.

Badham and H. B. Taylor, *Studies in Industrial Hygiene, no. 7, Joint Volumes of Papers Presented to the Legislative Council and Legislative Assembly, New South Wales*, vol. 1, 1st Session of the 28th Parliament, 1927, p. 52. Cited in B. Penrose, Occupational lead poisoning in battery workers: the failure to apply the precautionary principle, *Labour*

History, May 2003. Available at <http://www.historycooperative.org/journals/lab/84/penrose.html>.

Biomonitoring equivalents special issue, *Regul. Toxicol. Pharmacol.*, 2008, **51**, S3.

W. F. J. Busby and G. N. Wogan, Aflatoxins, in *Chemical Carcinogens*, ed. C Searle, American Chemical Society, Washington, D.C., 1984, p. 945.

S. M. Hays, R. A. Becker, H. W. Leung, L. L. Aylward and D. W. Pyatt, *Regul. Toxicol. Pharmacol.*, 2007, **47**, 96.

Health and Safety Executive, *Workplace Exposure Limits: Containing the List of Workplace Exposure Limits for use with the Control of Substances Hazardous to Health Regulations 2002 (as amended)*, HSE Books, London, 2005.

3.7 Local and Systemic Effects

local effect
Change occurring at the site of contact between an organism and a toxicant.
systemic
Relating to the body as a whole.
systemic effect
Consequence that is either of a generalized nature or that occurs at a site distant from the point of entry of a substance.

Note: A systemic effect requires absorption and distribution of the substance in the body.

Distinguishing between local and systemic effects is fundamental to the assessment of toxicological processes in poisoned organisms. Frequently, both types of effect are observed.

Local Effects. Local effects occur at the first site of contact on or in the body where application of the toxicant or exposure to it takes place. Examples are the immediate damage to the skin of contact with alkalis or acids, their similar corrosive effects on the intestine following ingestion or the direct effects of inhaled gases such as chlorine on the lungs. Chlorine causes swelling of the lung tissues, which may be fatal even if little or none is taken into the bloodstream.

Local effects generally occur quite rapidly after exposure, although consequences such as lung oedema may be prolonged, and they may precede systemic effects that then are of a different nature. An example is cadmium, which upon inhalation of high doses gives rise to lung oedema as an acute effect and renal tubular damage as a systemic effect. Thus, the identification of local effects can help to ensure rapid removal of anyone at risk from the exposure situation and also to provide prompt treatment of the intoxication, which may help to prevent systemic effects occurring. In some cases, the systemic effects observed may be secondary biological consequences of the local damage and not due directly to the harmful substance, *e.g.* kidney damage following severe acid destruction of the skin.

Systemic Effects. Systemic effects occur when a substance is absorbed, through the skin, from the gut, from the lungs, by injection, or is taken up by any other route, enters into the general blood circulation, is transported to various organs throughout the body and gives rise to effects on these organs. Once taken up in an organ, redistribution can occur by a release to the bloodstream and the agent can be taken up in another organ. This is seen for many metals, such as cadmium and lead. Most substances that are not highly reactive at body surfaces tend to produce systemic effects. Some substances produce both serious local and systemic effects. Tetraethyl-lead produces effects on skin at the site of absorption, which means that there is a local effect and then, after absorption, an effect on the central nervous system and other organs. Another example is phenol.

In general, while systemic toxicity may produce effects throughout the body, the major effects are on only one or more organs. These organs are referred to as target organs. However, the first organ that develops an adverse effect is called the critical organ for that particular toxicant. Many times, it is discussed whether the damage is reversible or irreversible. Although target organs show the most serious toxic effects, they are not necessarily the sites of highest accumulation of the toxicant. For example, lead is concentrated in bone, but its most serious effects are probably those on the brain. Common target organs are the brain and central nervous system, the circulatory system, the blood and haemopoietic system, the liver, kidneys, lungs and skin. Systemic effects often have target organ-specific names, *e.g.* neurotoxic (affecting the central nervous system), cardiotoxic (affecting the heart), hepatotoxic (affecting the liver) or nephrotoxic (affecting the kidney).

Increasing dose may permit a substance that causes a local effect to enter the affected organism and cause a systemic effect on a particular target organ. Increasing dose further will increase the number of target organs and effects until the whole organism is affected if it is not already dead.

Further Reading

A. W. Hayes, *Principles and Methods of Toxicology*, CRC Press, Boca Raton, FL, 5th edn, 2008.

3.8 Biomarker

biomarker
Indicator signalling an event or condition in a biological system or sample and giving a measure of exposure, effect, or susceptibility.

Note: Such an indicator may be a measurable chemical, biochemical, physiological, behavioural, or other alteration within an organism.

biomarker of effect
Biomarker that, depending on its magnitude, can be recognized as associated with an established or possible health impairment or disease.
biomarker of exposure
Biomarker that relates exposure to a xenobiotic to the levels of the substance or its metabolite, or of the product of an interaction between the substance and some target molecule or cell that can be measured in a compartment within an organism.
biomarker of susceptibility
Biomarker of an inherent or acquired ability of an organism to respond to exposure to a specific substance.

The initial changes in enzymes and other biological substances or physiological responses affected by a substance are called early effects, and some may be used as biomarkers of exposure and to give a measure of internal dose. The term 'biomarker' may cover any one of a range of biological effects reflecting the interaction between a toxicant and the organism affected. The term may be applied to a functional, biochemical or physiological change or it may be applied to a specific molecular interaction. The best biomarkers provide direct evidence for the exposure of individuals in a population to a particular substance, *e.g.* lead in bone, cadmium in the kidney (both *in vivo* determinations), mercury in urine or trichloroethylene in exhaled air. Quantitative measurements may permit the determination of a dose–effect relationship, particularly if the toxicokinetics of the substance are well established. Mostly, samples of blood, urine and exhaled air are used but hair, teeth and nail clippings may sometimes be analysed. Non-invasive methods should be used where possible. Such measurements may be used for screening or for monitoring either an individual or a group for absorbed dose.

Biomarkers of Exposure. Biomarkers of exposure, where applicable, are to be preferred to monitoring ambient media for exposure assessment because they reflect internal dose directly. An advantage of biomarkers of exposure is that they are an integrative measure, *i.e.* they provide information about exposure through all routes, including those of non-occupational exposure. An example where this is important is the combination of occupational exposure to lead with exposure to lead through hobbies (such as in soldering, shooting or glazing with lead) and with environmental exposure to leaded gasoline. Another example is occupational exposure to solvents combined with exposure to solvents at home during painting or while engaged in hobbies involving paint and glue.

Biomarkers related directly to exposure can be classified into two groups: (i) biomarkers of exposure (biological monitoring) and (ii) biomarkers of effect (biological effect monitoring) (Figure 22).

Use of Biomarkers of Exposure in Monitoring Individual Dose of Toxicants (Biological Monitoring). To assess internal dose and body burden, the amounts of toxicants of concern or their metabolites and/or derivatives in cells, tissues,

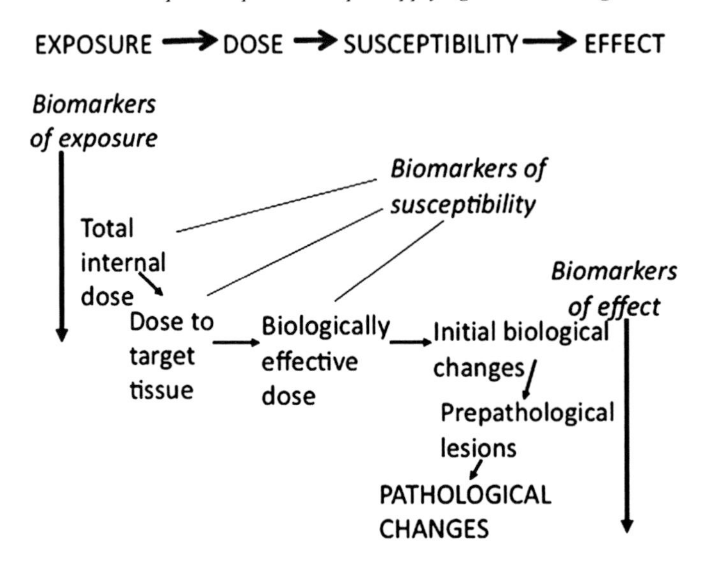

Figure 22 The different types of biomarker used in assessing exposure and its effects.

body fluids or excreta are measured. In addition or alternatively, an indirect biomarker of exposure may be determined such as cytogenetic change or reversible physiological change in exposed individuals. For example, if lead effects on haem synthesis are detected they may be used in addition to measurement of lead in blood as a measure of internal dose (integrated in time) for this particular effect. In the case of cadmium in blood, there is no effect on the blood and thus cadmium in blood has to be used exclusively as a measure of internal dose. An early effect of cadmium is on the kidney, causing the leakage of proteins into urine. A good dose measure to relate to this effect is the concentration of cadmium in urine, adjusted for urinary dilution. This measure also gives an indication of the body burden.

For assessment of internal exposure and dose, the sampling technique must be precisely controlled since it can profoundly affect results. Data for substances that disappear rapidly from the blood cannot be interpreted unless there is a documented standard time at which samples are taken. A standard time for sampling is especially essential if the half-life of the toxicant of concern in the body is short. Contamination is the major source of errors when analyzing many substances, especially metals such as nickel, chromium and cadmium. Contamination can come from the air, from the skin and sweat, sample containers and anticoagulants (for blood samples). The risk of contamination from skin, clothes and hair, as well as from the air at the workplace, is particularly great when collecting urine samples. Precipitation and adsorption are problems when collecting and storing urine samples. Certain chemicals, *e.g.* aluminium and volatile organic compounds, are adsorbed on glass and plastic.

The most common sample materials are anticoagulated whole blood, serum or plasma, urine and exhaled air. Saliva, sweat, hair and nails may also be used in biological monitoring for certain substances. Urine samples are frequently used as urine is easy to collect in large amounts. Variations in liquid intake and fluid loss (*e.g.* in a warm working environment where a lot of fluid is lost through sweat) result in large variations in concentrations of substances in urine. This variation is often corrected using the creatinine concentration of the urine or by measuring the urinary 24-h volume output. The relative density of urine can also be used for such corrections. Because of the differences between men and women, results for each gender should be reported separately.

Biomarkers of Effect (Biological Effect Monitoring). Biomarkers of effect are measurable biochemical, physiological or other alterations within an organism that can be recognized as associated with an established or potential health impairment or disease. Biomarkers of effect are often not specific for a certain substance but sometimes may be sufficiently specific to be used as surrogate measures of exposure and of dose. This will certainly be true if it is known that only one of the possible causes of a given effect can be involved in a given exposure situation. However, in some circumstances, single biomarkers of effect may be useful in blanket monitoring of multiple exposure, where those exposures have a common effect. If multiple biomarkers are used, and one or more markers are positive, additional markers or environmental monitoring can be used to determine the substance(s) causing the effect.

Examples of biomarkers of effect are:

1. The inhibition of certain enzymes of the haem synthesis pathway, which is caused by lead ions (or by dioxins), resulting in elevated concentrations of the precursors protoporphyrin and δ-aminolaevulinic acid dehydratase in blood and δ-aminolaevulinic acid and coproporphyrin in urine.
2. The leakage into urine of certain proteins such as β_2-microglobulin, α_1-microglobulin, retinol-binding protein and albumin, which is caused by several metal ions and solvents; in addition, there is inhibition of the activity of certain enzymes in the urine, *e.g.* *N*-acetyl-D-glycosaminidase (NAG), with specific isoforms of NAG-A and NAG-B.
3. The inhibition of the enzyme acetylcholinesterase, which occurs following exposure to several organophosphate and carbamate insecticides (*e.g.* parathion).
4. An increase in haemoglobin adducts, which follows exposure to aromatic amines, ethylene oxide, propylene oxide, butadiene and alkylating or arylating agents of all kinds. Such adducts may also be used as biomarkers of exposure.

Further Reading

IPCS, *Biomarkers in Risk Assessment: Validity and Validation*, EHC 222, WHO, Geneva, 2001. Available at <http://www.inchem.org/>.

3.9 Deterministic (Nonstochastic) and Stochastic Effects

stochastic
Pertaining to or arising from chance and hence obeying the laws of probability.
stochastic effect, stochastic process
Phenomenon pertaining to or arising from chance, and hence obeying the laws of probability.
deterministic effect, deterministic process nonstochastic effect.
nonstochastic process
Phenomenon committed to a particular outcome determined by fundamental physical principles.

It is important to distinguish between deterministic and stochastic effects because this determines how regulatory permissible exposure levels (PELs) are set, as discussed below.

Deterministic (Nonstochastic) Effects. Deterministic (nonstochastic) effects, also called threshold effects, are effects that have a threshold of chemical exposure below which they do not occur and above which the severity of the effect is related directly to the dose. Most toxic effects come into this category and, for these, determination of the threshold or some acceptable approximation to it (such as the benchmark dose described above) is the essential basis for setting the PELs that underlie all regulatory activities to protect people and their environment from harm that chemicals may cause.

Deterministic effects give the classical (or skewed) S-shaped dose–effect and dose–response curves, approximating to a linear relationship in the low dose area. The S-shaped curve is explained by the fact that both for receptor molecules and for individuals there is a range of sensitivity to any given toxicant. The variation in sensitivity is random (except in genetic extremes where a receptor or essential enzyme may be missing completely) and shows a normal (Gaussian) distribution.

If the population has genetic subgroups of markedly different susceptibility to the toxicant because of a missing or much altered receptor or activating enzyme, more than one such curve may be apparent in the test data, but this situation can be identified only if very large populations are studied.

Stochastic Effects. Stochastic effects, sometimes referred to as quantal effects, are usually produced by a reaction between an agent and DNA causing a discrete genetic change. Thus, they are described as 'all or none', either occurring or not occurring. For such reactions, no safe threshold of exposure can be determined, but the probability of their occurrence is related directly to increasing dose. Severity of the resultant toxic effect is not related to dose but to the consequences of the genetic change. The consequences of stochastic (or pseudo-stochastic) effects range from relatively minor metabolic deficiencies through immunologically mediated effects to mutagenicity, aspects of teratogenicity, genotoxic carcinogenicity and ultimately death. The inability to determine a threshold of exposure for stochastic effects does not mean that one does not exist, simply that it cannot be demonstrated by current methods of testing. In this circumstance, the precautionary principle determines that the no threshold assumption must be applied to ensure safety. As a result, PELs are set with reference to a calculated low probability of the adverse effect occurring. This probability for humans is usually set at 1 in 100 000 or at 1 in 1 000 000. To some extent, this is based on the practicality that an increase in disease rate (see below) of this order is too low to be detected by current techniques of epidemiological monitoring.

A problem with stochastic effects is that the shape of the dose–response curve is uncertain in at least three ways. First, the curves derived from animal trials are usually based on no more than two central data points (out of perhaps four doses used). Second, there is an inherent statistical uncertainty if an extrapolation is made to those low dose levels that are likely for normal human exposure and therefore of regulatory importance. Third, the curves are extrapolated to low doses using debatable mathematical models, usually chosen for precautionary purposes to give the highest estimate of probability of effect.

Further Reading

P. Illing, *Toxicity and Risk: Context, Principles and Practice*, Taylor and Francis, London; CRC Press, Boca Raton, FL, 2001.

3.10 Effect and Response

effect
Change in biochemistry, morphology, physiology, growth, development, or lifespan of an organism which results in impairment of functional capacity or impairment of capacity to compensate for additional stress or increase in susceptibility to other environmental influences.

response
Proportion of an exposed population with a defined effect or the proportion of a group of individuals that demonstrates a defined effect in a given time at a given dose rate.

In the general toxicological literature, the terms ‘effect’ and ‘response’ are often used interchangeably to describe a biological change in individuals or in a population that can be caused by a given exposure or dose. However, there is a clear distinction to be made between the consequences of exposure to an individual and those for a population. This is most clearly seen when one considers the differences in approach to human toxicology and ecotoxicology. In human toxicology, the main concern is protection or treatment of the individual, and the aim of regulatory activity is to protect every individual in the population at risk. In ecotoxicology, the aim is to protect populations, communities and ecosystems; the loss of a small number of individuals here is of little concern if the population can easily restore its numbers. Thus, it useful to be able to distinguish clearly between consequences for the individual and those for the population. This important distinction is facilitated if we differentiate between an effect and a response by applying the term 'effect' to a biological change in an individual and the term 'response' to the proportion of a population that demonstrates a defined effect. Following this convention, response means the incidence rate of an effect (Section 4.4). In this way, the LD_{50} may be described as the dose expected to cause a 50% response in a population tested for the lethal effect of a chemical.

It will be seen that the use of two different words distinguishing between effects on individuals and responses of populations makes for greater clarity of thought and communication. In human terms, effects on individuals must be understood for the application of toxicological knowledge to treat the problems of individual patients or to prescribe drugs properly. On the other hand, response of populations must be understood to regulate exposure to safe levels for both human populations and populations of other organisms.

Further Reading

E. J. Calabrese, L. A. Baldwin, Hormesis: the dose-response revolution, *Annu. Rev. Pharmacol. Toxicol.*, 2003, **43**, 175.

IPCS, Principles for modelling dose response for the risk assessment of chemicals, *Environmental Health Criteria Document Draft*, WHO, Geneva, 2004. Available at <http://www.who.int/ipcs/methods/harmonization/dose_response/en/>.

3.11 Teratogenicity

teratogenicity
Ability to cause the production of nonheritable structural malformations or defects in offspring.

Teratogenicity may be regarded as a special form of embryotoxicity or developmental toxicity. Unfortunately, the OECD Test Guidelines state that 'Developmental toxicology was formerly often referred to as teratology'. This statement is misleading since the term 'developmental toxicology' covers toxicity to the embryo and foetus in the widest sense. The distinctive definition of

Table 5 Teratogenic agents and their effects in humans.

Agent	*Effect*
Alkylating agents	Degeneration of the embryonic disc and intrauterine death
Androgenic hormones	Adrenal hyperplasia, virilization of the female genitalia
Chlorobiphenyls	Dark-brown staining of the skin, exophthalmos
Diethylstilbestrol	Virilization of the female foetus
Lithium	Cardiovascular defects
Retinoic acid	Multiple CNS and cardiac malformations, thymic aplasia
Thalidomide	Limb malformations, phocomelia, ear defects, *etc.*
Valproic acid	Neural tube defects, heart defects

'teratology' is required to concentrate attention on a special type of toxicity that requires a mechanism that affects the embryo and interferes with organogenesis, producing congenital malformations, *i.e.* irreversible functional or morphological defects present at birth. Malformations caused by exposures after birth are to be regarded separately as postnatal developmental disturbances. Table 5 lists typical teratogenic agents and the effects that they cause in humans.

Prevalence. Although there are many figures quoted for the prevalence of major congenital malformations, these figures are greatly dependent on the population studied, the point at which the data is collected after birth, and the classification of congenital defect (*e.g.* minor, major, cosmetic). Most sources use the figure of 3% prevalence rate for major congenital malformations recognized at birth, and another 3% rate for major congenital malformations unrecognized during the neonatal period. This combined 6% rate does not include mental and physical growth retardation, or minor congenital malformations such as hydrocele, angiomas, hernias and naevi that have not been regarded as serious enough for medical treatment. Of course, not all malformations can be attributed to drug use during gestation. The cause of approximately 40% of malformations is unknown. About 12–25% of these congenital malformations are purely genetic defects. Down's syndrome is the most common of this group. Another 20% are due to interactions between hereditary factors and environmental factors, the latter of which are largely unknown. Environmental factors alone, such as maternal disease or infection, chemicals and drugs, account for between 5% and 9% of the malformations. This category includes infections such as rubella (German measles), cytomegalovirus and *Toxoplasma gondii* (a protozoan) as definitive teratogens. It also includes maternal diseases such as diabetes and seizure disorders. Rubella infection is the most well known of the viral infectious teratogens. Maternal rubella can result in a group of defects, including heart disease, cataracts and deafness, known as foetal rubella syndrome. Diabetes is the most common chronic disease causing teratogenesis. Congenital malformations due to strictly environmental factors are estimated to occur in 0.1–0.2% of all live births. However, it is possible that accidental high exposures may lead to a much higher local incidence. Only a small portion of congenital malformations are due to drugs acting as teratogens.

Teratogenicity. Changes in embryonic tissues can be produced by chemicals at concentrations far below those causing target organ toxicity in adults. Further, teratogenic effects may occur in embryonic organs different from the target organs showing toxicity in adults. Possible effects in the offspring at birth or in the postnatal development period include embryo lethality (death), mild-to-severe dysmorphogenesis in one or more organ systems (resulting in structural malformations or physiological and biochemical dysfunction) and psychological, behavioural and cognitive deficits. Substances causing such effects are called teratogens. Teratogens are a major concern to the public, to industry and to regulatory agencies because of the low levels required to cause damage. Unfortunately, many Material Safety Data Sheets do not include much information on this increasingly recognized form of toxicity.

Timing is of the utmost importance in teratogenic studies, because exposure must occur during the period of organogenesis (Figure 23). Organogenesis occurs for different organs at different times after the implantation of the fertilized ovum. In other words, there are different windows of sensitivity for the different organs. In humans, organogenesis occurs in the first 8 weeks of pregnancy. Frequently, the period of chemical susceptibility is only a few days, with exposures before or after this time showing no effect upon a particular developing target organ, but perhaps causing adverse effects at another site. When using surrogate animal models to test teratogenicity, timing of

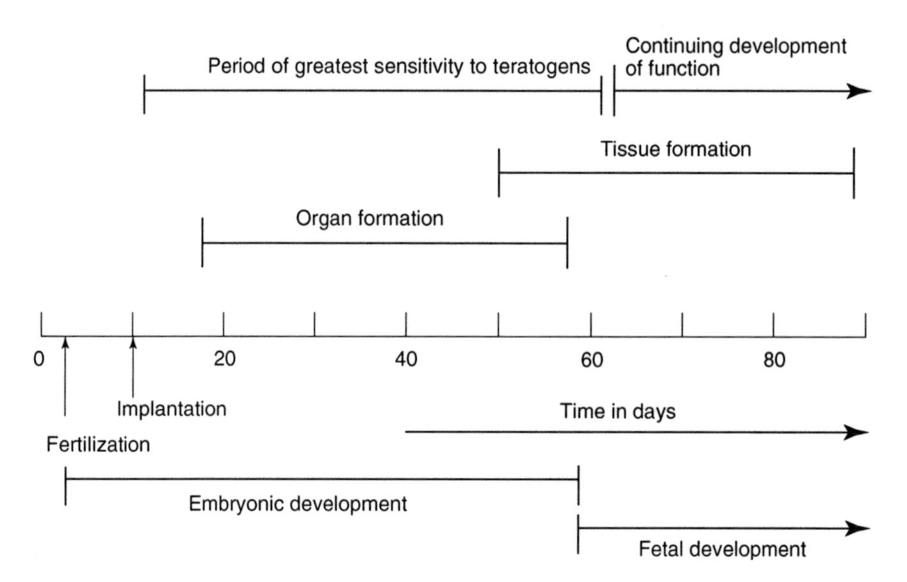

Figure 23 Stages in embryonic and foetal development showing the period of greatest sensitivity to teratogens.

administration is critical because of the shorter gestation time (*e.g.* 21 days in mouse and rat and 32 days in rabbit, as compared to 36 weeks in the human) and a corresponding reduction in the periods of organogenesis, often of only a few hours duration for certain organs. Because the sensitive stages of differentiation and development of embryos to the teratogenic effects of chemicals are often short in duration, the incidence of observable teratogenic effects of a given chemical may be rare in the potentially exposed population, as the combination of effective exposure and sensitive embryonic developmental stage may be a rare occurrence. This is a challenge both for the laboratory scientist and the epidemiologist trying to identify teratogens in order to regulate them.

Mechanisms of teratogenicity are not well understood and are probably of many different kinds, ranging from inactivation of key enzymes to small changes in hormonal balance at key stages in development. Oxidative stress and reactive oxygen species have been implicated in some teratogenic processes. Most recently, attention has focussed on the developing knowledge of epigenetics (Section 2.10) and inhibitions of histone acetylase or DNA methylation have been suggested as possible causes of teratogenic change.

Teratogenicity testing is routinely conducted in two species, a rodent (*e.g.* mouse or rat) and a non-rodent (usually the rabbit), and involves the daily administration of a range of three appropriate dose levels to different groups of timed-pregnant animals throughout the established period of organogenesis (*e.g.* days 6 to 15 for mice and rats or days 6 to18 for rabbits). The animals are killed 24 hours prior to the calculated day of parturition (day 20 for mice and rats, day 31 for rabbits) and undergo a complete necropsy. The number of live and dead foetuses is determined, and the uterine muscle is examined for re-absorptive sites (small scars) indicative of early embryonic deaths. Each foetus is weighed, the sex is determined and each is examined for external abnormalities (*e.g.* missing digits, limbs, tail; or eye, ear or cranial anomalies) prior to dissection to detect internal malformations (*e.g.* heart, lungs, intestinal tract, gonads). Whole foetuses are fixed in special fluids for histological examination of neural structures and for the detection of skeletal anomalies.

While the brain and peripheral nervous tissue begin to develop early in organogenesis, there is also considerable postnatal development that can be affected by prenatal exposure to potentially toxic agents, causing overt but subtle behavioural and cognitive deficits that cannot be detected by morphological examination of the brain. The possibility of such effects has resulted in modifications of the testing protocols, with some treated animals being allowed to give birth and to rear their young for 6 weeks after parturition (with or without continuous treatment of the dams). At 6 weeks of age, the young animals are examined by a battery of testing protocols designed to assess behavioural and cognitive development.

Test Protocols. There are two distinct sets of test protocols for reproductive toxicity including teratogenicity, one to be found in the OECD set of test guidelines, applied to industrial chemicals, and one established by the International Conference on Harmonization of Technical Requirements for Registration of Pharmaceuticals for Human Use (ICH). When interpreting data, it is important to know which of these has been used to obtain the data.

Further Reading

G. G. Briggs, R. K. Freeman and S. J. Yaffe, *Drugs in Pregnancy and Lactation*, Williams & Wilkins, Baltimore, 7th edn, 2005.

H. Kalter, *Teratology in the Twentieth Century*, Elsevier, Amsterdam, 2003.

International Conference on Harmonization of Technical Requirements for Registration of Pharmaceuticals for Human Use. Safety Guidelines, *Guidance on Specific Aspects of Regulatory Genotoxicity Tests for Pharmaceuticals – S2A*. Detection of Toxicity to Reproduction for Medicinal Products and Toxicity to Male Fertility – S5(R2). New Codification as per November 2005. Available at <http://www.ich.org/cache/compo/276-254-1.html>.

K. S. Korach, *Reproductive and Developmental Toxicology*, CRC Press, Boca Raton, FL, 1998.

OECD, *Guidelines for the Testing of Chemicals*, 2008. Available at <http://www.oecd.org/document_22/0,3343,en_2649_34377_1916054_1_1_1_1,00.html>.

WHO/IPCS, *Progeny, Principles for Evaluating Health Risks Associated with Exposure to Chemicals During Pregnancy*, EHC 30, World Health Organization, Geneva, 1984. Available at <http://www.inchem.org/>.

3.12 Reproductive Toxicity (See also Sections 3.11 and 4.3)

reproductive toxicant
Substance or preparation that produces nonheritable *adverse effects* on male and female reproductive function or capacity and on resultant progeny.

reproductive toxicity
Ability of a physical, chemical, or biological agent to induce adverse effects in the reproductive system.

reproductive toxicology
Study of the nonheritable *adverse effects* of substances on male and female reproductive function or capacity and on resultant progeny.

For purposes of classifying toxicities, the known induction of genetically based heritable effects in offspring is considered in Section 2.8 as a separate hazard class of germ-cell mutagenicity. This leaves the classification of reproductive toxicity to cover two groups of effects as in the *Globally Harmonized System of*

Classification and Labelling of Chemicals (GHS) and as proposed as working definitions in IPCS/EHC Document N° 225, *Principles for Evaluating Health Risks to Reproduction Associated with Exposure to Chemicals* (*see* Further Reading). The groups of effects covered are:

1. adverse effects on sexual function and fertility;
2. adverse effects on development of the offspring.

Adverse effects on sexual function and fertility may include, but are not be limited to, alterations to the female and male reproductive system, adverse effects on onset of puberty, gamete production and transport, reproductive cycle normality, sexual behaviour, fertility, parturition, pregnancy outcomes, premature reproductive senescence or modifications in other functions that are dependent on the integrity of the reproductive systems. Adverse effects on development, taken in the widest sense, include any effect that interferes with normal development of the conceptus, either before or after birth, and resulting from exposure of either parent prior to conception, or exposure of the developing offspring during prenatal development, or postnatally, up to the time of sexual maturation. However, classification under the heading of developmental toxicity is primarily intended to provide a hazard warning for pregnant women and for men and women of reproductive capacity. Thus, for regulatory classification, developmental toxicity means adverse effects induced during pregnancy, or as a result of parental exposure. The major manifestations of developmental toxicity include (i) death of the developing organism, (ii) structural abnormality, (iii) altered growth and (iv) functional deficiency.

Some reproductive toxic effects cannot be clearly assigned to either impairment of sexual function and fertility or to developmental toxicity. Nevertheless, substances with these effects would still be classified as reproductive toxicants. Adverse effects on or *via* lactation are also included in reproductive toxicity, but for regulatory classification such effects are treated separately. This is because it is desirable to be able to classify chemicals specifically for an adverse effect on lactation so that a specific hazard warning about this effect can be provided for lactating mothers.

Thus, from the above considerations, reproductive toxicants are classified into categories as follows:

- CATEGORY 1. Known or presumed human reproductive toxicant:
  This Category includes substances known to have produced an adverse effect on sexual function and fertility or on development in humans or for which there is evidence from animal studies, possibly supplemented with other information, to provide a strong presumption that the substance has the capacity to interfere with reproduction in humans. For regulatory purposes, a substance can be further distinguished on the basis of whether the evidence for classification is primarily from human data (Category 1A) or from animal data (Category 1B):

- CATEGORY 1A. Known human reproductive toxicant: Placing a substance in this category is based largely on evidence from humans.
- CATEGORY 1B. Presumed human reproductive toxicant: The placing of the substance in this category is based largely on evidence from experimental animals. Data from animal studies should provide clear evidence of an adverse effect on sexual function and fertility or on development in the absence of other toxic effects, or, if occurring together with other toxic effects, the adverse effect on reproduction is considered not to be a secondary non-specific consequence of these other toxic effects. However, when there is mechanistic information that raises doubt about the relevance of the effect for humans, classification in Category 2 may be more appropriate.
- CATEGORY 2. Suspected human reproductive toxicant:
  This category includes substances for which there is some evidence from humans or experimental animals, possibly supplemented with other information, of an adverse effect on sexual function and fertility or on development, in the absence of other toxic effects, or, if occurring together with other toxic effects, the adverse effect on reproduction is considered not to be a secondary non-specific consequence of the other toxic effects, and where the evidence is not sufficiently convincing to place the substance in Category 1. For instance, deficiencies in the study may make the quality of evidence less convincing, and, in view of this, Category 2 could be the more appropriate classification.

Effects On or Through Lactation. Effects on or through lactation are a separate toxicity classification. Substances that are absorbed by women and have been shown to interfere with lactation, or which may be present (including metabolites) in breast milk in amounts sufficient to cause concern for the health of a breastfed child, may be classified to indicate this. Classification can be assigned on the basis of:

1. absorption, metabolism, distribution and excretion studies that would indicate the likelihood the substance would be present in potentially toxic levels in breast milk; and (or)
2. results of one or two generation studies in animals that provide clear evidence of adverse effect in the offspring due to transfer in the milk or adverse effect on the quality of the milk; and (or)
3. human evidence indicating a hazard to babies during the lactation period.

Interpreting the Available Data on Reproductive Toxicity. Classification as a reproductive or developmental toxicant should be used for chemicals that have an intrinsic, specific property to produce an adverse effect on reproduction or development, and not if such an effect is produced solely as a non-specific secondary consequence of other toxic effects. In assessing toxic effects on developing offspring, it is important to consider the possible influence of maternal toxicity.

For human evidence to justify Category 1A classification there must be reliable evidence of an adverse effect on reproduction in humans. Evidence used for classification should be from well-conducted epidemiological studies that include the use of appropriate controls, balanced assessment and due consideration of bias or confounding factors. Less rigorous data from studies in humans should be supplemented with adequate data from studies in experimental animals and classification in Category 1B should be considered.

In some reproductive toxicity studies on experimental animals the only effects recorded may be considered to be of low or minimal toxicological significance and classification may not be justified. These include, for example, small changes in semen parameters or in the incidence of spontaneous defects in the foetus, small changes in the proportions of common foetal variants such as are observed in skeletal examinations or in foetal weights, and small differences in postnatal developmental assessments.

Data from animal studies should provide clear evidence of specific reproductive toxicity in the absence of other, systemic, toxic effects. However, if developmental toxicity occurs together with other toxic effects in the dam, the potential influence of the generalized adverse effects should be assessed. The preferred approach is to consider adverse effects in the embryo or foetus first, and then to evaluate maternal toxicity, along with any other factors that are likely to influence these effects. Developmental effects that are observed at maternally toxic doses should not be automatically discounted. Discounting developmental effects that are observed at maternally toxic doses can only be done on a case-by-case basis once a causal relationship is established or disproved.

If appropriate information is available it is important to try to determine whether developmental toxicity is due to a specific maternally mediated mechanism or to a non-specific secondary mechanism, like maternal stress and the disruption of homeostasis. Generally, the presence of maternal toxicity should not be used to negate findings of embryo or foetal effects, unless it can be clearly demonstrated that the effects are secondary non-specific effects. This is especially the case when the effects in the offspring are significant, *e.g.* irreversible effects such as structural malformations. In some situations it is reasonable to assume that reproductive toxicity is due to a secondary consequence of maternal toxicity and discount the effects, *e.g.* if the chemical is so toxic that dams fail to thrive, they are incapable of nursing pups, or they are prostrate or dying.

Fertility. Fertility includes aspects of spermatogenesis and oogenesis. Fecundity, fertility and 'sperm quality' are distinct parameters that are not equivalent and are frequently confused. Sperm count and sperm quality do not necessarily predict whether conception will take place for a given couple. A 'fertile couple' is one that has conceived at least one child. Fecundity is the ability of a couple to conceive a child and is often evaluated by the time necessary to achieve pregnancy [time to pregnancy (TTP)].

Infertility may be thought of as a negative outcome. A couple that is infertile is unable to have children within a specific time frame. Therefore, the epidemiological measurement of reduced fertility or fecundity is indirect and is accomplished by comparing birth rates or time intervals between births or pregnancies. This may be done by several methods, including the standardized birth ratio [(SBR; also referred to as the standardized fertility ratio) and the length of time to pregnancy or to birth (TTP)]. In these evaluations, the couple's joint ability to produce children is estimated. The SBR compares the number of births observed with those expected based on the person-years of observation, preferably stratified by factors such as time period, age, race, marital status, parity and contraceptive use. The SBR is analogous to the standardized mortality ratio, a measure frequently used in studies of occupational cohorts, and has similar limitations in interpretation. The SBR is obviously less sensitive in identifying an effect than semen analyses.

'Time to pregnancy' is also a useful tool and has clearly demonstrated a difference in fecundity among smokers and non-smokers. Analysis of the time between recognized pregnancies or live births is a more recent approach to indirect measurement of fertility. Because the time between births increases with increasing parity, comparisons within birth order (parity) are more appropriate.

A specific effect that may reduce fertility is damage to sperm chromatin. This is assessed by the sperm chromatin structure assay (SCSA). SCSA is a flow cytometric technique that identifies the spermatozoa with abnormal chromatin packaging, defined by susceptibility to acid denaturation *in situ*. Acridine orange staining is used, after a low pH challenge, to distinguish between denatured (red fluorescence = single stranded) and native (green fluorescence = double stranded) DNA regions in sperm chromatin, the former being a result of DNA breaks and (or) of changes in protamine quantity and composition and (or) of an insufficient level of disulfide groups. The level of DNA breaks is conveniently expressed by the DNA fragmentation index (DFI), which is the ratio of red to total (red plus green) fluorescence intensities in the flow cytometric analysis. In two independent studies, DFI levels $>30 \pm 12$ were incompatible with fertility *in vivo* of otherwise normal sperm.

In reproductive toxicology, it is important to be aware of factors such as the genotype of the foetus and possible interactions between genes and the environment. Most knowledge is based on animal data from laboratory animals. The differences in reproductive toxicity between species are exemplified by the consequences of exposure to the drug thalidomide. Humans and monkeys are very sensitive to the effects of thalidomide while mice and rats are not. Thus, tests on mice and rats indicated that thalidomide had low toxicity and failed to identify its toxic effects on the foetus. The ability to cause gross structural (anatomical) malformations in the developing embryo or foetus exemplified by the effects of thalidomide is called teratogenicity (Section 3.11).

By the time metabolic pathways and excretion are fully developed in the newborn, the probability of adverse effects may decrease for some toxicants and increase for others. The kidney does not have full function before delivery and is not fully functional during the immediate postnatal period. This leads to limited excretion of potential toxicants *via* the urine. Metabolism and excretion of some potentially toxic substances like caffeine, for example, is very limited in the foetus. An example of the consequences of deficient metabolism in the foetus is the grey baby syndrome seen in newborn infants. In 1959 this syndrome was reported in association with the use of the antibiotic chloramphenicol. Infants developed abdominal distension, vomiting, cyanosis, cardiovascular collapse, irregular respiration and subsequent death shortly after therapy with chloramphenicol was started. Pharmacokinetic studies in the neonate showed accumulation of chloramphenicol in plasma due to poor drug metabolism. In other cases when bioactivation of xenobiotics can take place to form active metabolites, toxicity may also be increased.

It is very important to be aware of differences between individuals and between species. Studies combining molecular biology with classical epidemiological approaches have demonstrated the existence of allelic variants for developmentally important genes that may enhance the susceptibility of the embryo. For example, the association between heavy maternal cigarette smoking (>10 cigarettes a day) and cleft lip and (or) palate in the offspring is marginally significant until an allelic variant for tumour growth factor (TGF-) is considered. The combination of smoking and the uncommon variant for the gene raises the odds ratio to a highly significant level. With regard to species variation, endocrine disruption and subsequent reproductive toxicity have been identified as an effect of environmental chemicals, such as DDT and PCBs, that can cause feminizing effects in the population (Section 4.3).

Historically, reproductive toxicity in relation to development has concentrated on the effects of exposure of females. However, it is now clear that exposure of males to certain agents can adversely affect their offspring and cause infertility and cancer. The hazards associated with exposure to ionizing radiation have been recognized for nearly a century and there is evidence for harmful effects of paternal exposure from X-ray studies in mice resulting in heritable tumours. In humans, smoking fathers appear to give rise to tumours in the F1 generation, *i.e.* the first filial generation produced by two parents. Cyclophosphamide, 1,3-butadiene and urethane have been tested in rodents and, after exposure of males only, each compound produced an adverse effect in F1 male offspring.

To protect the public from the harmful effects of reproductive toxicants, labelling must include appropriate warnings. Directives have already been published by the EU covering about 70 CMR (carcinogenic, mutagenic and reprotoxic) substances. Recent changes to the cosmetics laws in some countries have banned the use of CMR substances in cosmetics. There is continuing revision of CMR classification because of the international movement, promoted by the United Nations, towards a Globally Harmonized System of Classification and Labelling of Chemicals (GHS).

Further Reading

IPCS, *Principles For Evaluating Health Risks To Reproduction Associated With Exposure To Chemicals*, EHC 225, WHO, Geneva, 2001. Available at <http://www.inchem.org/>.

P. Apostoli, S. Telisman and P. R. Sager, Reproductive and developmental toxicity of metals, in *Handbook on the Toxicology of Metals*, ed. G. F. Nordberg, B. A. Fowler, M. Nordberg and L. T. Friberg, Academic Press Elsevier, Amsterdam, 3rd edn, 2007, pp. 213–249.

D. Anderson, Male-mediated developmental toxicity, *Toxicol. Appl. Pharmacol.*, 2005, **207,** S506.

UNECE, *Globally Harmonized System of Classification and Labelling of Chemicals (GHS)*, UNECE, Geneva, 2nd revised edition, 2007. Available at <http://www.unece.org/trans/danger/publi/ghs/ghs_rev02/02files_e.html>.

3.13 Immunotoxicity, Immunosuppression and Hypersensitivity

hypersensitivity
State in which an individual reacts with allergic effects following exposure to a certain substance (allergen) after having been exposed previously to the same substance.

Note: Most common chemical-induced allergies are type I (IgE-mediated) or type IV (cell-mediated) hypersensitivity.

immunosuppression
Reduction in the functional capacity of the immune response; may be due to:
1 Inhibition of the normal response of the immune system to an antigen.
2 Prevention, by chemical or biological means, of the production of an antibody to an antigen by inhibition of the processes of transcription, translation or formation of tertiary structure.

immunotoxicity
Ability of a physical, chemical, or biological agent to induce adverse effects in the immune system.

There are many possible toxic effects on the immune system, and in a broad sense they may be classified as either immunosuppression or unwanted immune activation. In broadest terms, immunosuppression leads to a deficit in the immune response to microbial pathogens and transformed cells, whereas inappropriate activation is associated with allergy, hypersensitivity and autoimmunity. The immune system plays a key part in protecting us from infectious disease by attacking and destroying infectious organisms and potentially harmful macromolecules. It also monitors our own cells and attacks and destroys many of those that become transformed to potentially cancerous progeny. This host resistance to infectious agents and neoplasms depends on the presence of immunocompetent cells that may be stimulated to proliferate in the presence of antigenic substances.

The immune system depends for its efficient function on the intrinsic cooperation of various cells, which is strongly dependent upon the biological actions of soluble mediating molecules such as immunoglobulins, hormones, growth factors and cytokines acting through membrane receptors. Any substance that affects these interactions can cause agent-specific or species-specific damage that, in many cases, causes immunosuppression (resulting in, for example, decreased resistance to infectious agents and development of tumours). Immunosuppression may be either systemic or occur at the local level (*e.g.* in lungs or skin). As noted above, an inappropriate increase in the immune response can also occur, leading to hypersensitivity, as exemplified by respiratory tract allergy or allergic contact dermatitis. Furthermore, some substances can cause the development of autoimmune diseases.

Diseases associated with abnormal immune function, including common infectious diseases and asthma, are more prevalent at younger ages. Several factors may explain this increased susceptibility, including functional immaturity of the immune system and age-related differences in metabolism. Although not yet conclusively demonstrated, it is generally believed that the immature immune system is more susceptible to xenobiotics than the fully mature system, and that the effects of immunotoxicant exposure in the very young may be particularly persistent. This contrasts with effects observed following exposure in adults, which generally occur at higher doses and do not persist for long after exposure ends. Experimental animal studies show that harmful effects on the developing immune system may be qualitative (*e.g.* affecting the developing immune system without affecting the adult immune system) or quantitative (*e.g.* affecting the developing immune system at lower doses than in adults). Immune maturation may simply be delayed by xenobiotic exposure and may recover to normal adult levels over time. However, if exposure interferes with a critical step in the maturation process, lifelong defects in immune function may follow. These defects may be expressed as immunosuppression or as dysregulation of the immune system, resulting in decreased resistance to infection or development of a functional phenotype that is associated with allergy and asthma. Experimental evidence indicates that development can be hindered or delayed to such an extent that certain effector mechanisms either are absent or do not function properly for essentially the lifetime of the individual, as has been reported following exposure to diethylstilbestrol (DES) or tetrachlorodibenzo-*p*-dioxin (TCDD). In humans, the clinical effects of immunotoxicant exposure during development may be expressed immediately or later in life, presenting as either increased infectious or neoplastic diseases, or as increased incidences (or severity) of allergic or autoimmune disease.

The complexity of the immune system results in there being many potential target sites and pathological consequences of toxicity. The strategies devised by immunotoxicologists working in safety assessment have all applied a tiered panel of assays to identify immunosuppressive and immunostimulatory agents in laboratory animals. These testing strategies include measurement of one or more of the following: altered lymphoid organ weights and histology; changes in the cells of lymphoid tissue, peripheral blood leukocytes and (or) bone marrow; impairment of cell function at effector or regulator level; and altered susceptibility to challenge with infectious agents or tumour cells. With respect to clinical evaluation and research on human responses for immunoactivation, the lymphocyte transformation (proliferation) test performed on peripheral blood lymphocytes, cytokine profiling of blood or serum, and analysis of lymphocyte subpopulations by flow cytometry are currently under scrutiny.

Assessment of Immunotoxicity. Many factors must be considered in evaluating the potential of an environmental agent or drug to influence adversely the immune system of experimental animals and humans. As for assessment of any type of toxicity, these of course include selection of appropriate animal models and exposure variables, an understanding of the biological relevance of the end-points being measured, use of validated measures and quality assurance. The experimental conditions should take into account the potential route and level of human exposure and any available information on toxicodynamics and toxicokinetics. The doses and sample sizes should be selected so as to generate clear dose–response curves, needed for the determination of NOAEL or NOEL. Testing should be continually refined to allow better prediction of conditions that may lead to disease. In addition, techniques should be developed that will help to identify mechanisms of action; these may include methods for *in vitro* examination of local immune responses (such as in the skin, lung and intestines), and techniques of molecular biology and the study of genetically modified animals. These obvious criteria require special attention in assessing immunotoxicity, where appropriate measurements and end-points are not always clearly agreed.

Unfortunately, detection of immune changes after exposure to potentially immunotoxic substances is much more difficult in humans than in experimental animals. Testing possibilities are limited, levels of exposure to the agent (*i.e.* dose) are often difficult to establish, and the immune status of populations is extremely heterogeneous. Age, race, sex, pregnancy, acute stress and the ability to cope with stress, coexistent disease and infections, nutritional status, tobacco smoke and medications may all contribute to this heterogeneity. This heterogeneity renders less effective some of the clinical tests mentioned above.

Since most of our knowledge regarding human toxicity of environmental chemicals comes from epidemiological studies, epidemiological study design for such investigations must be appropriate if we are to reach correct conclusions. The commonest design used in studies of immunotoxicity is the cross-sectional study, in which exposure status and disease status are measured at one time or over a short period. The immune function of 'exposed' subjects is then compared with that of a comparable group of 'unexposed' individuals. Because many of the immune changes seen in humans after exposure to a chemical may be sporadic and subtle, populations should be studied soon after exposure and sensitive tests must be used for assessing the immune system. Conclusions about immunotoxic effects should not be based on changes in a single parameter, but in the immune profile of an individual or population.

Examples of Immunotoxicants. Examples of environmental and industrial chemicals reported to be immunotoxicants in humans are asbestos, benzene and halogenated aromatic hydrocarbons, including 2,3,7,8-tetrachlordibenzo-*p*-dioxin (TCDD; often referred to just as dioxin) and polychlorinated biphenyls (PCBs), which give rise to immunosuppression. TCDD and PCBs are widespread environmental pollutants that are resistant to biodegradation and have lipophilic properties facilitating bioaccumulation in, for example, fish, and thus in humans who eat the fish. Other economically important chemicals known to be powerful hypersensitizers are the isocyanates. In studies of the toxicity of metallic elements, a dose–response relationship is sometimes seen that shows stimulation of some immune functions at low doses while high doses cause immunosuppression. Nickel ions and chromium ions cause hypersensitivity reactions (see below) and dermatitis in humans. Beryllium is another metal with hypersensitizing and immunotoxic properties.

Hypersensitivity. When exposed to an antigen, the body may produce antibodies specific to that antigen. These antibodies may provide immunity against later exposures to the antigen. Under some physiological conditions, or in people with defective immune systems, an excessive immune reaction may occur that can cause cell and tissue damage. Histamines released from mast cells following such damage can cause dilation of small blood vessels, tissue inflammation and constriction of the bronchi of the lungs. The result may be anaphylaxis, an immediate, sometimes fatal, hypersensitivity reaction to some substances, including macromolecules occurring in the diet, notably to some found in peanuts. In humans, the clinical signs and symptoms of anaphylaxis include reaction of the skin with itching, erythema and urticaria; reaction of the upper respiratory tract with oedema of the larynx; reaction of the lower respiratory tract with dyspnoea, wheezing and cough; reaction of the gastrointestinal tract with abdominal cramps, nausea, vomiting and diarrhoea; and reaction of the cardiovascular system with hypotension and shock. Individuals undergoing anaphylactic reactions may develop any one, a combination, or all of the signs and symptoms. Anaphylaxis may be fatal within minutes, or fatality may occur days or weeks after the reaction. Death is a result of the damage suffered as a result of the decrease in blood pressure following extreme dilation of the blood vessels.

Serum sickness is a similar but milder hypersensitivity to serum proteins or drugs that occurs several weeks after injection of foreign material. Delayed reaction hypersensitivity occurs when lymphocytes react to certain antigens. The lymphocytes slowly infiltrate an area, such as skin exposed to poison ivy toxin, and cause local inflammation reactions that may be followed (or accompanied by) tissue damage. Anaphylaxis, serum sickness and delayed sensitivity may occur in otherwise normal individuals as well as those inclined to allergies. Individuals with allergic, or atopic, hypersensitivity form antibodies that react with antigens to cause local tissue damage and such symptoms as hives, hay fever

Table 6 Types of hypersensitivity.

Type	*Alternative name*	*Related health problems*	*Mediators*
1	Allergy	Atopy Anaphylaxis Asthma	IgE
2	Cytotoxic, antibody dependent	Erythroblastosis fetalis Goodpasture's syndrome Autoimmune haemolytic anaemia	IgM IgG Complement
3	Immune complex disease	Serum sickness Arthus reaction Systemic lupus erythematosus (SLE)	IgG Complement
4	Cell-mediated	Delayed hypersensitivity Contact dermatitis Tuberculosis Chronic transplant rejection	Lymphocytes

and asthma. Antihistamines are drugs that prevent histamine from acting on blood vessels, bronchioles and other organs. Acute reactions, such as anaphylaxis, are treated by giving epinephrine and other sympathomimetic drugs to support the blood circulation. Steroids such as cortisone are also given to suppress inflammation and depress the immune system. In some cases, hypersensitized individuals receive injections of gradually increasing quantities of the antigenic material to which they are sensitive, to build tolerance and avoid or lessen their hypersensitivity to that particular substance.

Hypersensitivity has been classified into four types by Gell and Coombs (*see* Further Reading) (Table 6).

Autoimmunity. Autoimmunity is a condition in which the immune system fails to differentiate self-antigens from foreign antigens and begins to attack self-tissues. In effect, the body generates an immune response against itself. An autoimmune reaction results in inflammation and in tissue damage. Some of the more common autoimmune disorders are rheumatoid arthritis, systemic lupus erythematosus, and vasculitis. It is also possible that some glomerulonephritides, Addison's disease, mixed connective tissue disease, polymyositis, Sjögren's syndrome and progressive systemic sclerosis, have autoimmune components. It is likely that some cases of infertility may be explained by autoimmunity. Some common multifactorial diseases have an autoimmune component, and Type I diabetes is an important example. Common antigens in generalized autoimmunity are the nuclear antigen recognized by the anti-nuclear antibody that is important in the diagnosis of systemic lupus erythematosus, and the anti-mitochondrial antibody elevated in primary biliary cirrhosis. Some autoimmune diseases are poorly understood but may, in certain circumstances, be a consequence of exposure to a toxic substance or substances. For example, haemolytic anaemia has been linked to an autoimmune response affecting the red cells following exposure to the pesticide dieldrin.

Skin Sensitivity. Allergic contact dermatitis (ACD) is a major cause of minor discomfort to people. It is also the most tested form of immunotoxicity and is largely a preventable disease. The effects can be prevented by correct identification of skin sensitizers, characterization of their potency, understanding human skin exposure and application of good risk assessment and management strategies.

Skin sensitization is caused by substances that behave as electrophiles and can react with skin proteins, altering them so that the immune system recognizes them as 'foreign'. This leads to proliferation of lymphocytes that recognize the substance and (or) the altered protein that it produces. Subsequent contact with the substance can then cause an enhanced effect on the skin, producing redness, swelling, itching, *etc*. Substances that do this are called skin sensitizers. People who are sufficiently exposed may develop the sensitized state called contact allergy; once they have acquired that state, further exposure can lead to the development of allergic contact dermatitis.

For assessing the risk of developing ACD from exposure to a given substance, it is important to know the relative potency of an identified chemical allergen, to understand how people are exposed to it, and to integrate this information into the risk assessment. Having done this, one can formulate a plan for management and reduction of the risk. For consumers, this may mean voluntary or regulatory imposition of concentration limits in products, often with suitable labelling. Control of occupational exposure may involve both an equivalent of the consumer limit process and a greater emphasis on controlling skin exposure, if necessary by the use of personal protective equipment.

Multiple Chemical Sensitivity (MCS). Multiple chemical sensitivity (MCS) is essentially an adverse reaction of susceptible persons to a large number of chemicals. When exposed to the relevant chemicals, people with MCS react with symptoms such as nausea, headache, dizziness, fatigue, impaired memory, rash, and respiratory difficulty. Many household and industrial chemicals, including cleaning products, tobacco smoke, perfumes, inks and pesticides, have been mentioned as triggers for MCS. It can be argued that immunotoxicity provides a mechanism for some of the observed symptoms but there is no conclusive evidence of this.

Most toxicologists and physicians do not regard MCS as a legitimate medical syndrome, arguing that the depression that frequently accompanies it is an indication that the symptoms are psychosomatic. Further, descriptions of the syndrome are largely anecdotal and difficult to verify scientifically. It is also true that the imprecisely defined syndrome is easily misused as a diagnosis, leading to a large number of worker's compensation cases involving MCS. Nevertheless, many putative sufferers do seem to improve when they eliminate contact with the chemicals suspected of triggering their condition; in extreme cases, this seems to require confinement to specially treated living quarters.

Further Reading

Encyclopedia of Immunology, ed. P. J. Delves and I. M. Roitt, Academic Press, London, 2nd edn, 1998. Available at <http://www.roitt.com/>.

Clinical Aspects of Immunology, ed. P. G. H. Gell and R. R. A. Coombs, Blackwell, Oxford, 1st edn, 1963.

3.14 Elimination and Clearance

clearance (in toxicology)
1 Volume of blood or plasma or mass of an organ effectively cleared of a substance by elimination (metabolism and excretion) divided by time of elimination.

> *Note*: Total clearance is the sum of the clearances of each eliminating organ or tissue for a given substance.

2 (in pulmonary toxicology) Volume or mass of lung cleared divided by time of elimination; used qualitatively to describe removal of any inhaled substance which deposits on the lining surface of the lung.
3 (in renal toxicology) Quantification of the removal of a substance by the kidneys by the processes of filtration and secretion; clearance is calculated by relating the rate of renal excretion to the plasma concentration.

elimination (in toxicology)
Disappearance of a substance from an organism or a part thereof, by processes of metabolism, secretion, or excretion.

elimination rate
Derivative with respect to time of the concentration or amount of a substance in the body, or a part thereof, resulting from elimination.

'Clearance', and the related term 'elimination', are used in toxicology in an attempt to describe, usually quantitatively, the rate of disappearance of a substance (drug, toxin, analyte, *etc.*) from an organism, organ, tissue or compartment. 'Elimination' refers simply to the disappearance of the substance. 'Elimination rate' indicates the time over which this happens, and so is a derivative with respect to time (dc/dt) of the concentration (c, expressed in suitable units) of the substance in the compartment of interest. The use of these terms does not imply, or make any reference to, mechanism. The substance may be removed, or the compartment cleared, by mechanical means such as transport or filtration, or by conversion into another substance through metabolism or biotransformation.

Clearance. Whereas elimination focuses on the substance and its rate of change of concentration with time, clearance generally refers to the compartment, such as blood or lung, in the numerator. (Renal clearance is somewhat different and is discussed below.) It is common to report clearance from blood (or from plasma if the substance does not enter the blood cells) as this is a measure of how long cells and tissues perfused by the circulation will be exposed to the substance. The numerator of the term, then, would be blood volume, and we would say for instance that so many millilitres of blood were cleared of the substance in a minute. This is quite different from an elimination rate where the expression is dc/dt. We do not mean that the substance is completely eliminated from the reference volume of blood in a given time. That is, we do not focus on 1 mL of blood and say that c drops to zero. Rather, if the concentration drops by one-half we would say that this has removed enough substance to clear one-half of the blood volume.

One reason it is useful to express clearance in this way is that it relates to the total volume of the compartment, *e.g.* blood. So if elimination is due to the functioning of a particular organ, for example, this gives us an indication of the organ's functional capacity. It also allows for consideration of changes in blood volume, for instance, due to transfusion or haemorrhage. This means of expression is also mathematically useful in compartment models.

A note is added to the definition of 'clearance' that total clearance of a substance is the sum of individual clearances from each eliminating organ or tissue. This might seem intuitively obvious from conservation of mass. But it reminds us that (i) we may not always be able to identify all sources of elimination, (ii) in multi-compartment models, subtraction from one compartment may result in addition to another and (iii) this relation holds whether concentration of substance or volume of compartment is being divided by time.

Clearance as commonly used in pulmonary toxicology has a somewhat different nuance. When a substance is inhaled and deposits on the epithelial surface of the bronchi, bronchioles or alveoli, clearance is expressed as disappearance from this surface. The value is only qualitative, for two reasons: (i) as we cannot easily measure the surface area, we use lung mass or volume instead, but of course the surface-to-volume ratio is not known; (ii) we cannot really measure surface concentrations. The clearance will be derived by considering the amount of surface deposition that could arise from a certain exposure, and the time it takes to drop to a presumed value of zero, determined by analytical detection, with inherent detection limits, following a procedure such as lavage. Because we define clearance in pulmonary toxicology on the basis of tissue surface, elimination may be either mechanical, through mucociliary transport and (or) the cough reflex, or by cellular uptake. When the substance is taken up by the cells, it may remain in the lung tissue or be eliminated by metabolism or absorption into the body, but it has been cleared from the lining surface of the respiratory tissues.

Clearance in renal toxicology has a somewhat specialized meaning, reflecting the unique physiology of the kidney. Renal clearance is quantitative. While it is often more useful to know the effect of removal a substance from the body fluids than to know the composition of the urine, plasma clearance frequently depends primarily on renal clearance, and the latter is an important measure of renal function. A substance with a molecular weight (relative molar mass) of less than about 65 kDa will be filtered from human plasma at the renal glomerulus, and then reabsorbed (or not) to varying degrees during subsequent passage through the remainder of the nephron. For a substance that is removed from plasma exclusively by urinary excretion, its concentration in urine divided by its concentration in plasma will equal its plasma clearance divided by the urinary rate of flow. With units as examples:

$$[\text{concentration in urine/mol L}^{-1}]/[\text{concentration in plasma/mol L}^{-1}] = [\text{plasma clearance/mL min}^{-1}]/[\text{urine flow/mL min}^{-1}]$$

The glomerular filtration rate (GFR) is a measure of glomerular function. If a substance is cleared from the plasma solely by the kidney and is not reabsorbed or secreted, then the GFR is equal to the plasma clearance. Inulin is a substance that behaves this way to a very good approximation, and its plasma clearance can be used to measure GFR. Creatinine is an endogenous substance produced by muscle metabolism at a fairly constant rate and shows a reasonable approximation to inulin with respect to its handling by the body. Thus, creatinine clearance is a good clinical measure of GFR. Creatinine is easier to measure than inulin and, as an endogenous substance, its use is less invasive. Glucose is completely reabsorbed in normoglycaemic states. Thus, in renal physiology glucose clearance is often said to be zero, although it is of course non-zero in more general terms because it is taken up by cells and metabolized. Because the kidney can also secrete substances through the peritubular capillaries independently of glomerular filtration, plasma clearance can be greater than GFR. An important example of this latter principle is H^+. The opposite is true of a substance that is reabsorbed by the kidney tubules.

Further Reading

D. J. Greenblatt, Elimination half-life of drugs: value and limitations, *Annu. Rev. Med.*, 1985, **36**, 421.

Concept Group 4. Concepts Applying to Environmental Toxicology

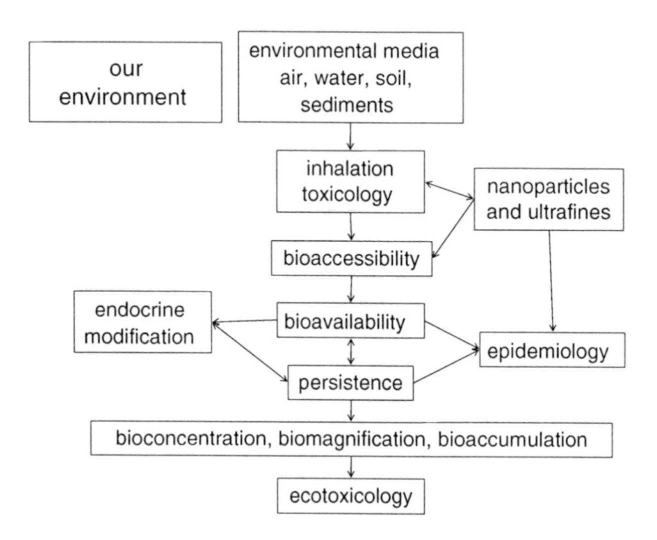

Concepts in Toxicology
By John H Duffus, Douglas M Templeton and Monica Nordberg

Published by the Royal Society of Chemistry, www.rsc.org

4.1 Persistence

persistence
Attribute of a substance that describes the length of time that the substance remains in a particular environment before it is physically removed or chemically or biologically transformed.

This section concentrates on substances that have the potential to be either water or fat soluble or may become so as a result of chemical reactions in the environment or of biotransformation by living organisms. Persistence is also associated with insoluble particulates (usually dusts, *e.g.* silica, or fibres, *e.g.* asbestos) that may have adverse effects following deposition in the lung and inadequate removal by lung clearance mechanisms. Such particulates may also cause harm following reactions with the gut and, if small enough (Section 4.2), with other organs after absorption into the body. In non-mammalian organisms, they may interact with gills and body or cell surfaces to produce adverse effects. They may be phagocytosed with consequences for both prey and predators. They may trap nutrients, restricting their bioavailability. They may act as catalysts for reactions that may be biologically harmful or beneficial. To consider these possibilities properly requires a more extensive discussion than is possible here.

Persistence of potentially water-soluble or fat-soluble substances reflects the ability of such substances to remain in an environment unchanged, or changed to another potentially toxic form, for a long period of time, sometimes for many years. The longer a substance persists, the greater the potential for exposure of humans and other organisms, and for bioaccumulation to toxic concentrations. The persistence of a substance is usually measured or estimated as a half-life in air, water, soil and (or) sediment. The persistence of a substance in air is important because, once in the atmosphere, it may travel global distances from its original point of release. Persistence in air is a property of volatile persistent organic pollutants (POPs). POPs have been defined as organic chemicals that are stable in the environment, are liable to long-range transport, may bioaccumulate in human and animal tissue, and may have significant impacts on both human health and the environment. Such chemicals are resistant to hydrolysis and to breakdown by the action of ultraviolet light and oxygen. They include compounds such as dichlorodiphenyltrichloroethane (DDT), furans, polychlorinated biphenyls (PCBs) and dioxins.

It is important to distinguish between persistence in a single medium and overall environmental persistence. Persistence in an individual medium is controlled by transport of the substance to other media, as well as by transformation into other chemical species. Together, these factors determine the residence time in the medium. Persistence in the environment as a whole is a distinct concept. The environment behaves as a web of interconnected media, and any substance released to the environment will become distributed in these media in accordance with its intrinsic reactivity and physicochemical properties.

Multimedia (multi-compartment) mass balance models are the most convenient means for determining the overall environmental persistence from information on sources and loadings, chemical properties and transformation processes, and inter-media partitioning. Notably, these models reflect an equilibrium or steady-state situation within a defined geographic space (which is why these models are sometimes referred to as box models). On the global scale, an equilibrium situation will never be reached and a steady-state situation is not likely.

Multimedia mass balance models may be subdivided into three categories on the basis of the level of complexity addressed:

Level I – Predicts the equilibrium distribution of a fixed quantity of conserved chemical, in a closed environment at equilibrium, with no degrading reactions, no advective processes and no inter-media transport processes (*e.g.* no wet deposition or sedimentation).

Level II – Predicts the result of continuous discharge of a substance at a constant rate into the total environmental system until steady-state and equilibrium conditions are reached at which the input and output rates are equal. Loss of the substance is assumed to occur by degradation reactions and by advection. Inter-media transport processes (*e.g.* wet deposition or sedimentation) are not quantified.

Level III – Predicts the result of a chemical being continuously discharged at constant rates independently into each environmental medium such as air, water and soil, until it achieves a steady state in which input and output rates are equal. Loss is assumed to be by degradation reactions and advection. Unlike the Level II model, occurrence of equilibria amongst media is not assumed and, in general, each medium is at a different fugacity. A mass balance applies not only to the system as a whole, but also to each compartment. Rates of inter-media transport are calculated using so-called D values introduced by Donald Mackay (see Mackay, 2001, in Further Reading) to derive expressions for transport between media (*e.g.* evaporation) or for transformation (*e.g.* biodegradation). D values are expressed in mol h^{-1} Pa^{-1} and contain information on mass transfer coefficients, areas, deposition and resuspension rates, diffusion rates and soil runoff rates. When a substance present in a phase is subject to several transport and transformation processes, each has its own D value and the relative importance of each process is made clear by the magnitude of the D value. Note that, unlike the assumptions in Level II models, inputs to each medium in Level III models are assumed to occur independently. In Level II only the total input into the whole system is considered.

An estimate of overall environmental persistence may be obtained using a Level III Mackay fugacity model from the Canadian Environmental Modeling Centre website (*see* Further Reading) for a standardized set of environmental conditions in which net advection and sediment burial are ignored. The purpose of such a model is to provide the user with a measure of overall environmental persistence based on chemical reactivity only. This may be useful in making decisions for the purpose of pollution prevention because ultimately, on a global scale, there is no advective removal. If advective loss is included, the residence time is reduced and may give a misleading impression of a short persistence. Advective losses merely relocate the chemical; they do not destroy it.

It is important to note the difference between residence time and half-life in each environmental medium when interpreting the overall persistence. The half-life is a measure of how fast a chemical is eliminated from each environmental medium by chemical transformation or degradation. The residence time in each medium includes the half-life but it also takes into account transport into and out of that medium. The residence time may, therefore, be substantially different from the media-specific half-life for some chemicals.

Bioaccessibility, Bioavailability, Bioaccumulation, Bioconcentration and Biomagnification. These terms are discussed in other sections of this book but must be referred to here because of their central importance in relation to the consequences of persistence of POPs and similar substances. POPs, if they are bioaccessible, have the physicochemical properties, notably lipid solubility and octanol–water partition coefficient (Section 3.2), required to be readily bioavailable. The same properties facilitate uptake by living organisms (bioaccumulation and bioconcentration) and, consequently, biomagnification through food webs, to potentially toxic levels in predators.

Toxicity. Toxicity is a function of the concentration of the chemical and the duration of exposure. Because persistent and bioaccumulative chemicals are long-lasting substances that can build up in the food chain to high levels they have a higher potential to express toxicity and to be harmful to humans and the ecosystem.

Media. There are four major environmental compartments in which a chemical may be found: air, water, soil and sediment. Air is also referred to as the atmosphere or atmospheric compartment. The atmospheric compartment includes the area closest to the Earth's surface (the troposphere) and beyond (the stratosphere). Water refers to surface water, and includes oceans, rivers, lakes and ponds. This definition does not include aquifers found below the Earth's surface (groundwater).

Soil refers to the topmost layer (the first few inches) of the Earth's surface that is not covered by water. Soil includes all the material near the surface that differs from the underlying rock material. Sediment is the particulate matter found under surface water. Usually soil is found to be aerobic (oxygenated) and sediment is anaerobic (free of oxygen). However, soil can become anaerobic and sediment may be aerated following dredging. This can have a profound effect on persistence of substances.

Half-life. Half-life is the length of time it takes for the concentration of a substance to be reduced by one-half relative to its initial level, assuming first-order decay kinetics. It is a useful approximation to consider 'complete removal' as taking approximately six half-lives, *i.e.* $1/2^6 = 1/64$ of the amount of substance remains. This may be quite misleading for highly toxic substances.

Laboratory-based half-lives for any given substance are generally related only to reactivity in a given defined medium under one or a few sets of defined conditions. Laboratory studies may also fail to replicate the microbial potential for biodegradation of an environment at risk.

Percentage in Each Medium. Once a substance is released into the environment, it may move from one environmental compartment to another. For example, volatile substances deposited on land will move into the air. Some substances such as DDT or dichlorodiphenyldichloroethylene (DDE) enter the air as vapour in hot climates and then condense in the cold air of the Arctic and Antarctic, or in the snow at the top of high mountains. Water-soluble substances will end up in rivers and the sea. The expected distribution of a substance in air, water, soil and sediment is expressed as the estimated percentage in each medium relative to the total amount in the environment. The amount in each medium at steady state may be calculated using the Mackay fugacity models available freely from the website of the Canadian Environmental Modeling Centre (see Further Reading below).

Further Reading

Canadian Environmental Modeling Centre, 2008. Fugacity Models available at <http://www.trentu.ca/academic/aminss/envmodel/models/models.html>.

European Commission, *Technical Guidance Document on Risk Assessment in Support of Commission Directive 93/67/EEC on Risk Assessment for New Notified Substances and Commission Regulation (EC) No 1488/94 on Risk Assessment for Existing Substances and Directive 98/8/EC of the European Parliament and of The Council Concerning the Placing of Biocidal Products on the Market*, 4 volumes, European Chemicals Bureau, Ispra, 2006. Available at <http://ecb.jrc.ec.europa.eu/home.php?CONTENU=/DOCUMENTS/TECHNICAL_GUIDANCE_DOCUMENT/EDITION_2/>.

D. Mackay, *Multimedia Environmental Models: The Fugacity Approach*, CRC, Boca Raton, FL, 2nd edn, 2001.

D. Mackay and E. Webster, Environmental persistence of chemicals, *Environ. Sci. Pollut. Res.*, 2006, **13**, 43.

4.2 Nanoparticles and Ultrafine Particles

nanoparticle
Microscopic particle whose size is measured in nanometres, often restricted to so-called nano-sized particles (NSPs; < 100 nm in aerodynamic diameter), also called ultrafine particles.

ultrafine particle
Particle in air of aerodynamic diameter < 0.1 μm.

Note: The fraction of the air particulates referred to as ultrafines is often referred to as the $PM_{0.1}$ fraction.

The term, 'nanotechnology' was introduced in 1974 by Norio Taniguchi (*see* Further Reading) to describe new technology for designing semiconductors on an atom-by-atom or molecule-by-molecule, basis. In essence, nanotechnology deals with formation and usage of small structures that have useful properties and can be manipulated at the atomic or molecular level.

The term 'nanoparticles' originated in relation to nanotechnology to describe the synthetic particles that make it possible. Nanoparticles are increasingly used in body lotion, facial creams and sun screen protection; in food stuffs; and in a great variety of other products used by humans. Previously, the term 'ultrafine particles' was applied to particles in this size range of <100 nm in aerodynamic diameter, referring to naturally formed particles found in the air. It is difficult to see any toxicological value in distinguishing between nanoparticles and ultrafine particles on the basis that one group is synthetic and the other natural. This situation mirrors the distinction commonly made between natural and synthetic chemicals, a distinction that has limited toxicological significance.

Ultrafine particles (and nanoparticles) can be absorbed in the lungs, mostly from the alveoli. The uptake in an organism of particles *via* the lungs reflects the balance between deposition and clearance. Deposition results from the impaction of large particles and the sedimentation and diffusion of small particles. Total deposition is a measure of all the particles deposited. For particles of aerodynamic diameter of about 3 μm, deposition is found to be approximately 50%. Larger particles are deposited in the upper respiratory tract, while smaller particles deposit in the lower airways and alveoli. Clearance of larger particles from the bronchial tree is by mucociliary transport into the pharynx. This typically takes about 24 hours from the time of deposition to clearance, but depends on status of the mucociliary clearance. Transport of smaller particles by macrophages from the alveoli may take much longer.

A large percentage of ultrafine (and nano-) particles are exhaled, *i.e.* are breathed out again, although a significant number of inhaled particles are deposited in the respiratory tract. In addition, there is evidence of uptake of these very small particles into the skin and the olfactory system, as well as through the gut wall following ingestion. Uptake through the skin may require special attention because of the large surface area available for absorption. Uptake through the olfactory mucosa may be followed by transport through nerves of the olfactory system into the brain. There are medical uses of nanoparticles that may entail their injection.

Chemistry of Nanoparticles. The term 'nanoparticles', as mentioned above, has been applied particularly to man-made ultrafine particles containing, for example, titanium oxide, zinc oxide, silica and carbon. Sometimes these are called 'engineered nanoparticles', implying nanoparticles manufactured to have definite properties or a definite composition. The surface properties of nanoparticles are likely to be important in determining their toxicity. New and improved techniques will be required to characterize them and consequently their toxicologically significant properties.

There has been particular concern about nanotubes that consist of graphene sheets that form single-walled (SWNT) or multi-walled (MWNT) nanotubes. Single-walled carbon nanotubes (SWCNT) have unique properties: they are stiffer than diamond, and are estimated to be ten-times stronger than steel. In the preparation of carbon nanotubes (CNTs), metals, amorphous carbon and other compounds are used. Nanotubes may include catalytically active metals like nickel and iron. Both nickel and iron catalyse oxidative reactions. Ultrafine particles (and nanoparticles) can adsorb lipopolysaccharides and cause pro-inflammatory effects. Contamination of gold nanoparticles with lipopolysaccharides can interfere with the immune system.

Risk Assessment of Nanoparticles. Respirable nanoparticles may be more toxic than larger particles of equivalent mass and of the same composition. Thus, characterizing the exposure to such particles involves more than simply analysing the content of molecules or elements. It is unclear to what extent macrophages can phagocytose different kinds of ultrafine or nanoparticles of different types, sizes and shapes. Nanoparticles have been reported to contribute to development of coronary heart disease, but again the harm of specific nanoparticles has not been well characterized. Asbestos fibres represent naturally occurring ultrafine particles of well-established toxicity and carcinogenicity. Their length and diameter are crucial to their toxic effects, and similar defined dimensions may determine the toxicity of nanoparticles. It may be that the surface area of nanoparticles is the best measure of dose, but the lesson learned from the extensive characterization of the health effects of asbestos fibres is that the three-dimensional shape of nanoparticles will likely affect their toxicity.

Whether nanoparticles can generally cause cancer remains to be established. Some metal compounds that are used for making nanoparticles, *e.g.* titanium oxide, have been classified as potential human carcinogens, and nanoparticles incorporating such metallic elements will require further careful assessment from the regulatory point of view.

Further Considerations. The study of the toxic effects caused to living systems, particularly human beings, by exposure to nanoparticles brings together chemistry, bioengineering and toxicology to predict harmful effects that may occur as exposures of humans and their environment increase. The aim of nanotoxicology is to define the steps required to avoid such effects. There is an immediate problem in defining appropriate units of dose and exposure. Mass or substance measurement is not appropriate since it does not correlate in any simple way to the number of particles or their surface area, both of which are likely to be important factors in biological interactions. Surface shape is also important in any interaction of particulates with living cells and tissues. Such interaction involves surface chemistry and electric charge, and so these become further considerations in nanotoxicology. After interaction with the body surface, wherever this may occur, further reactions will be affected by the structure of the nanoparticles, including such physical characteristics as crystallinity and porosity. Put another way, after the initial interaction, distribution of nanoparticles in the body of an organism will be a function of both the surface characteristics of the particles and the surface characteristics of different parts of the body. In addition, a critical size may exist beyond which the movement of nanoparticles is restricted. Thus, dose–effect and dose–response studies may require to be stratified according to defined particle characteristics, including mass, number, size, shape, surface area, crystallinity and, if known, mechanism of action.

The above factors have not yet been well characterized. There will surely be characteristics of nanoparticles that will target them to specific organs, tissues and cells; these remain to be elucidated. The presence of components such as metals present in nanotubes may determine their health effects. An increased risk of cardiopulmonary diseases associated with nanoparticle exposure requires special attention, if it is not to become a serious problem. This raises questions about the ethical justification of experiments that have been carried out with volunteers to study the effects of ultrafine particles and that have demonstrated harmful effects, even if apparently minor.

Conclusions. Although there may be common risks associated with all exposures to nanoparticles, each type of nanoparticle must be treated individually for the assessment of health risks. Traditional toxicological testing is, in general, inappropriate when carrying out risk assessments for nanoparticles. The general properties of nanoparticles as a group require the design and validation of a set of special toxicity tests for adequate risk assessment. Nanoparticles designed for drug delivery or as food components should be given particular attention in this regard.

Further Reading

P. J. A. Borm, D. Robbins, S. Haubold, T. Kuhlbusch, H. Fissan, K. Donaldson, R. Schins, V. Stone, W. Kreyling, J. Lademann, J. Krutmann, D. Warheit and E. Oberdorster, The potential risks of nanomaterials: a review carried out for ECETOC, *Particle Fibre Toxicol.*, 2006, **3**, 11. Available at <http://www.particleandfibretoxicology.com/content/3/1/11>.

K. Donaldson, L. Tran, L. Jimenez, R. Duffin, D. E. Newby, N. Mills, *et al.*, Combustion-derived nanoparticles: A review of their toxicology following inhalation exposure, *Particle Fibre Toxicol.*, **2**, 2005, 10.

Available at <http://www.particleandfibretoxicology.com/content/2/1/10>.

N. Taniguchi, *Proc. Intl. Conf. Prod. London, Part II*, British Society of Precision Engineering, 1974.

4.3 Endocrine Modification

endocrine disruptor
endocrine modifier
Exogenous chemical that alters function (s) of the endocrine system and consequently causes adverse health effects in an intact organism, its progeny or (sub)populations.

In 1962, Rachel Carson pointed out in her book *Silent Spring* (*see* Further Reading) that synthetic chemicals, such as dioxins, dichlorodiphenyltrichloroethane (DDT) and polychlorinated biphenyls (PCBs), were affecting the fertility, reproductive success and behaviour of wild animals. This implied that there was interference with their endocrine systems.

Evidence for similar human health problems came from the work of Herbst, Ulfelder and Poskanzer (*see* Further Reading), who observed that treatment of women during the 1950s and 1960s with diethylstilbesterol (DES), a synthetic estrogen agonist, had resulted in a marked increase in vaginal cancer among their daughters. This was supported by further studies by other workers and these led to a book entitled *Estrogens in the Environment*, edited by John McLachlan (*see* Further Reading). The effects of DES may be an example of the epigenetic effects described in Section 8.

In 1993, Richard Sharpe in the UK and Nils Skakkebaek in Denmark reported epidemiological studies apparently showing decreasing sperm count and sperm motility among men who were born after the Second World War in Denmark and Scotland. In a hypothesis paper in the *Lancet* (*see* Further Reading) they advanced the idea that environmental chemicals could be a cause for this decline in male fertility. Since then, increased rates of testicular cancer, undescended testes and hypospadias have been attributed to endocrine disruption by synthetic environmental agents. Other examples of endocrine disruption (or modification) may be decreasing age at menarche, decreasing male-to-female sex ratio at birth, and various congenital malformations. The ultimate result could be a reduction in reproductive potential. It has been suggested that the observed increased incidence of neurobehavioral disorders and cancer of organs under hormonal influence may also be a consequence of exposure to endocrine disrupting chemicals. Even the current increasing incidence of obesity has been suggested to result from the ability of some endocrine disruptor (ED) substances to trigger fat-cell activity. However, the epidemiological data for effects of EDs have been questioned and no environmental ED has been clearly identified as a possible cause of specific effects.

Endocrine Disruption in Wildlife. In 1991, an interdisciplinary conference considered the proposition that synthetic chemicals released into the environment could affect endocrine systems and alter the reproduction, development and physiology of wildlife and humans. During the conference, the terms 'endocrine disruption' and 'endocrine disruptor' were coined. The proceedings of the conference were published in 1992 as a book, *Chemically-Induced Alterations in Sexual Development: The Wildlife/Human Connection* (*see* Further Reading). The conclusions were that some environmental chemicals, described as 'endocrine disruptors', may act as hormone agonists or antagonists, or may interfere with hormone synthesis, and thus disrupt endocrine networks in animals and humans. It was postulated that embryonic and young organisms might be particularly sensitive to such disruptions. This includes adolescent organisms during sexual maturation, a process that depends on the correct levels and timing of hormone production. Thus, exposing the embryo, foetus, or even the maturing child or adult to endocrine disruptors at biologically significant doses could potentially cause many diseases, ranging from cancer or physical malformations to immunological and neurological disorders and infertility.

Compared with humans, there is much more evidence that wildlife has been affected adversely by exposures to EDs, starting with the evidence of poor reproduction of DDT and PCB contaminated birds, cited by Rachel Carson, and caused by eggshell thinning and abnormal behaviour. Subsequently, there has been evidence of feminization of alligators exposed to DDT and dicofol in contaminated lake water, and of feminization of fish exposed to human female hormones and residues from hormonal drugs. Even more striking was the imposex effect of tributyltin oxide on dog whelks, causing females to develop male genital apparatus, making them sterile after exposure to concentrations barely detectable in seawater.

The strength of the evidence for endocrine disruption in wildlife may reflect the fact that many studies have been conducted in areas where the levels of environmental chemicals are high (*e.g.* where there are point source discharges, such as occur in the Great Lakes and the Baltic Sea). These studies have mostly concentrated on animals inhabiting aquatic ecosystems, which bioaccumulate certain EDs. In contrast with human studies, it has also been possible to experiment with the animal species of concern under both laboratory and field conditions. However, there are problems in determining the full range of potential effects of EDs on wildlife because of the large number of potential target species, differences in physiological mechanisms, and lack of detailed knowledge of endocrine function in many species.

Chemical Structure of Endocrine Disruptors. The chemical structure of potential EDs varies enormously. Examples of such chemicals are synthetic and natural hormones, plant estrogens, some pesticides, some persistent organic pollutants (POPs), bisphenol A, various metal species and certain phthalate esters. Both phthalates and bisphenol A are high-production volume chemicals. In general, these substances have estrogenic effects although some phthalates studied are anti-androgenic and some may have anti-thyroid action. Polybrominated diphenylethers (PBDEs), which are used as flame retardants, have also been reported to interfere with thyroid metabolism.

Bisphenol A is of particular concern because it is a structural analogue of DES. It is widely used as an antioxidant in plastics and as an inhibitor in polymerization of poly(vinyl chloride)s. Bisphenol A has been associated with developmental toxicity, carcinogenic effects and possibly neurotoxicity, in addition to its estrogenic effect. It is found in plastic tubes and syringes and in pacifiers for babies. It leaches from plastic bottles. Some countries have banned selling baby products that might contain bisphenol A because of its perceived potential to cause harm to infants and children.

Endocrine disruptors may affect the gonadal-pituitary axis and modify the metabolism and (or) synthesis of estradiol and testosterone. Examples are the reductase inhibitors that block production of dihydrotestosterone. Such inhibitors are used as drugs, but are also found in herbs such as saw palmetto. Several chemicals are estrogen mimics that bind to the high-affinity estrogen-binding protein called the estrogen receptor (ER). These chemicals include drugs, natural products and manufactured chemicals. It has been postulated that such estrogen mimics may be involved in causing breast cancer, uterine cancer and developmental defects. There is also concern about herbal remedies that are marketed for 'female' and 'male' health. These preparations are physiologically active and modify the target endocrine system.

Substituted phenols such as bisphenol A, discussed above, and the longer chain nonyl- and octyl-phenols are widespread in the environment but it should be emphasized that they are many times weaker than the active human hormone estradiol in standard assays. Several derivatives of the pesticides DDT and methoxychlor are active as estrogens and (or) anti-androgens through mechanisms that involve the respective estrogen or androgen receptors. Some natural products, such as genistein from soy, exhibit remarkable estrogenic activity.

Conclusions. ED research has tended to focus on compounds that persist and bioaccumulate in exposed organisms. Considering the variations in the endocrine system during development and aging, and in response to environmental conditions, there is a need to pay more attention to the consequences of timing, frequency and duration of exposure to potential ED substances. It must also be noted that there is a wide range of possible mechanisms for endocrine disruption, depending on the substances involved and the animal species that is exposed. Endocrine disruption itself is not a precisely defined toxicological end-point and subsidiary end-points must be identified. These may range from interaction with molecular receptors up to the level of physiological or anatomical changes. While EDs have been recognized to be of particular concern in the regulatory context, no experimental strategy to identify them has been generally agreed and validated.

Whether the substances defined as EDs are indeed a serious threat to the environment or public health is not clear. Many of the human diseases and disorders predicted to increase as a result of low environmental exposure to EDs have not yet done so. However, the ED hypothesis has drawn attention to the challenge of understanding the complexity of hormone action. Many studies are underway on breast cancer, endometriosis, testicular cancer and other plausible end points but it must be remembered that these illnesses may have multiple causes and may be influenced by individual life styles and habits. Thus, the idea that endocrine disruption in humans is caused by environmental chemicals is still a hypothesis requiring more evidence before it can be considered proven.

Further Reading

R. Carson, *Silent Spring*, Boston University Press, Boston, Massachusetts, 1962.

Chemically-Induced Alterations in Sexual and Functional Development: The Wildlife/Human Connection, ed. T. Colborn and C. Clement, Princeton Scientific Publishing Co., Princeton, NJ, 1992.

A. Herbst, H. Ulfelder and D. Poskanzer, *New Engl. J. Med.*, **284**, 1971, 878.

Estrogens in the Environment, ed. J. A. McLachlan, Elsevier, New York, 1979.

R. M. Sharpe and N. E. Skakkebaek, *Lancet*, 1993, **341**, 1392.

S. Sathyanarayana, *Curr. Probl. Pediatr. Adolesc. Health Care*, 2008, **38**, 34.

US EPA, *Endocrine Disruptor Screening Program (EDSP)*, 2008. Available at <http://www.epa.gov/scipoly/oscpendo/pubs/edspoverview/index.htm>.

4.4 Rate in Toxicology

rate (in epidemiology)
Measure of the frequency with which an event occurs in a defined population in a specified period of time.

Note 1: Most such rates are ratios, calculated by dividing a numerator, *e.g.* the number of deaths, or newly occurring cases of a disease in a given period, by a denominator, *e.g.* the average population during that period.

Note 2: Some rates are proportions, *i.e.* the numerator is contained within the denominator.

The rate of a chemical reaction is a straightforward concept and refers to the amount of substance that reacts, *i.e.* is converted into another chemical species, in a given period of time. Such rates are preferentially expressed in molar units divided by time. Thus, the IUPAC 'Gold Book' defines rate as a derived quantity in which time is a denominator quantity, adding that the rate of x is dx/dt. A rate constant in chemical kinetics is, then, a proportionality constant that adds the dimension of time to the relationship in concentration between two species. This fairly obvious point is made because we will distinguish two different uses of rate that are important in toxicology and toxicokinetics. One meaning that is consistent with the time base refers to the rate of transfer of a substance between different compartments, pools, sources or sinks with time as the denominator. Formally, we can treat a substance with the same mathematical formalism whether it is reacting chemically to produce a new product or being shuttled between compartments. Another meaning of rate is perhaps less obvious to the chemist: in epidemiology, a rate is more generally a frequency where the denominator may be a population. In the following paragraphs, we introduce several important rates that can be discussed based on intercompartment trafficking. Then we return to the idea of rate as a frequency.

Rates of Importance in Toxicology. A discussion of some important rates could begin with mention of the rate of uptake from external media. Factors affecting the uptake of a substance into the body invoke discussion of bioavailability, and then the substance is transported (described by formal rates in compartmental analysis), metabolized [metabolic rate, and see biotransformation (Section 2.4) or bioconversion] or eliminated [compare with rates of clearance or elimination (Section 3.14)]. Regarding clearance, the GFR (glomerular filtration rate) is an important concept. It refers to the volume of serum ultrafiltrate cleared of a substance by passage through the glomerular capillaries, again per time.

Clearance from a compartment may reflect more generally metabolism, excretion or transfer to another compartment. For instance, the rate of clearance of a reactive organic intermediate would contribute to the clearance of the substance. But, we also speak of a rate of biotransformation. This concept is closest to chemical terminology, where the biotransformation usually reflects an enzymatically catalysed phase I or phase II reaction. The rate here is a clear chemical rate constant, where the mechanism is known or presumed.

The somewhat different use of rate in epidemiology requires further explanation. We often think of a rate as a change with time, and the concept of a rate as a proportionality factor is not intuitively obvious. In its first use, the term is closely connected with the chemical (arithmetical) concept of the rate constant. When we add the qualification that 'the numerator is contained within the denominator' we move out of the strict dimension of time to allow a rate or proportionality relative to another denominator. For example, prevalence 'rate' in epidemiology is the total number of all individuals who have an attribute or disease at a particular time divided by the population at risk of having the attribute or disease at this point in time. Thus, the numerator, the number of individuals with the attribute, is included in the denominator, the total population at risk. In this epidemiological usage, we are calculating a proportion and not a temporal rate in the physicochemical sense.

Rate is defined in the *Shorter Oxford English Dictionary* as

> a stated numerical proportion between two sets of things (the second usually expressed as unity), especially as a measure of amount or degree

(*moving at a rate of 50 miles per hour*) or as the basis of calculating an amount or value (*rate of taxation*) or rapidity of movement or change (*travelling at a great rate*).

In the past, a mathematical derivative was described in words as, *e.g.* the *rate* of change of y with respect to x, with no implication that time has to be involved. We can also say that the Gaussian probability distribution decreases at a rapid rate on each side of its maximum, among many other examples. 'Rate' in chemistry should always be coupled with 'of reaction' or 'of diffusion', *etc.* to indicate that we refer explicitly to a particular time dependence.

Use of the word 'frequency' in the definition of the epidemiological sense of rate again introduces the concept of time, though in a different sense as, here, frequency involves a certain number of occurrences in a fixed period of time. We would speak of the number of events (*e.g.* deaths) from a particular cause, based on the whole population, as a rate. Here, the denominator would logically be a number of people, and the resulting rate would be a dimensionless proportionality. The rate, then, is the number of affected or identified individuals divided by the whole population. We define 'incidence' as the number of new individuals succumbing to a particular event or illness in a period of time. Prevalence is the number of events existing per unit population at a given time.

So, in epidemiological terms, both incidence and prevalence refer to time as a base, and thus are rates, though they express somewhat different concepts. Incidence is the number of new cases, *e.g.* of individuals falling ill, normalized by population, in a period of time, whereas prevalence is defined as the number of incidences of a disease or other events existing at a given point in time. Rates are affected by environmental and physiological conditions. Knowledge of rates of exchange among body compartments is the key to producing effective toxicokinetic models. Knowledge of rates in epidemiology as proportions is central to developing environmental models that are needed for risk assessment.

Further Reading

J. M. Last, *A Dictionary of Epidemiology*, Oxford University Press, Oxford, 4th edn, 2001.

M. M. Szklo and F. J. Nieto, *Epidemiology: Beyond the Basics*, Jones and Bartlett, Boston, 2nd edn, 2007.

4.5 Ecotoxicology

ecotoxicology
Study of the toxic effects of chemical and physical agents on all living organisms, especially on populations and communities within defined ecosystems; it includes transfer pathways of these agents and their interactions with the environment.

Toxicology originated as a science concerned mainly with the effects of toxicants on humans. Ecotoxicology was developed from toxicology in the late twentieth century following the realization that pollution of the natural environment was having effects on other organisms, and that consequential effects were threatening human welfare.

Ecotoxicology can be considered at three levels. First, there is direct toxicity to an individual species; human toxicology is an example of this. Second, there are toxic effects on inter-relationships between species, *e.g.* the effect of excess nutrients in causing eutrophication, algal blooms, release of toxins and subsequent anoxia in affected waters. Third, there is accumulation of toxicants by individual organisms and their movement between organisms and species through predator/prey relationships that results in biomagnification, *e.g.* the transfer of persistent organic pollutants like organochlorines from water to birds of prey, adversely affecting both their reproduction and tolerance of stress with a resultant decline in numbers. Underlying these considerations is the movement of potentially toxic substances through the natural environment and their physicochemical transformations in the different environmental media.

Ecotoxicology incorporates basic concepts of ecology, allowing the observed toxic effects to be interpreted, predicted and prevented. Furthermore, ecotoxicology encompasses knowledge both of how organisms interact in nature with each other (the biotic environment) and of the physical and chemical aspects of the environment (the abiotic environment). In this, ecotoxicology differs from human toxicology that concentrates on effects of substances on individuals. Even when human populations are considered through epidemiology, the aim is normally to obtain knowledge of cause and effect to protect individuals at risk. In contrast, ecotoxicology is used to protect populations, species and communities.

Ecosystems. Many environmental conditions must be satisfied for life on Earth to be possible but, quantitatively, there are two major requirements that all organisms have to sustain life. The first is a supply of carbon to form the organic molecules of which organisms are composed. The second is a supply of energy for the chemical reactions that keep the organisms alive and particularly for the biosynthetic processes that maintain their structure and function. Carbon is freely available in the environment as carbon dioxide in the air, as various inorganic forms, including carbonate and bicarbonate, and as organic carbon. With regard to organic carbon, organisms can be divided into two major groups, heterotrophs and autotrophs. Heterotrophs are organisms with a requirement for presynthesized organic molecules. Animals, fungi and most bacteria are heterotrophs. Autotrophs are organisms that are independent of outside sources for organic food materials and manufacture their own organic material from inorganic sources. Green plants are autotrophs that use light energy to trap carbon and convert it into complex organic derivatives.

Clearly, toxicity to autotrophs must have an adverse effect on the heterotrophs that depend upon them, simply by depriving them of nutrition. On the other hand, some autotrophs, *e.g.* some algae, synthesize and release toxins. If such algae flourish, any heterotrophs that can eat them safely will grow in numbers while others may be eliminated by the toxins released. Hence, differential sensitivity to toxicants may result in changes in species balance that may destabilize an ecosystem. Such considerations of differential species sensitivity to environmental factors and of species interdependence are characteristic of ecotoxicology.

Energy and Carbon. The number of autotroph species is relatively small but they are essential components of ecosystems because they make the organic compounds on which all living organisms depend. Most autotrophs are green plants. Fixation takes place by photosynthesis. The fundamental chemistry of this can be represented simply as follows:

$$6CO_2 + 6H_2O + h\nu \rightarrow C_6H_{12}O_6 + 6O_2$$

The term $h\nu$ represents photon energy, and photosynthesis is countered by respiration, which can be represented thus:

$$C_6H_{12}O_6 + 6O_2 \rightarrow 6CO_2 + 6H_2O + \text{chemical energy}$$

The chemical energy in the respiration equation is available for use in the cell in the form of ATP, now placing the energy term on the right-hand side of the equation. The energy released by respiration (a catabolic reaction and part of catabolism) is used for biosynthesis of components of the organism (anabolism) and for other life processes.

Only autotrophs make new organic matter while all organisms use or consume it. Fixation of carbon by autotrophs is called primary production. Subsequent incorporation of already-existing organic matter into heterotrophs is called secondary production. Thus, primary production by autotrophs must be sufficient to meet the needs of both autotrophs and heterotrophs. In a balanced ecosystem, there is a balance between production and catabolism. This means that total respiration should be balanced by photosynthesis to maintain the optimum carbon dioxide level in the atmosphere. It appears that, at the time of writing, this balance is no longer being maintained and that carbon dioxide is building up in the atmosphere and contributing to global warming. Cutting carbon dioxide production will help to correct this, but attention must also be paid to maintaining autotrophs and their photosynthetic activity.

Other Nutrients. In addition to energy and carbon, living organisms need at least 20 different elements in appropriate chemical species – for physiological processes, biochemical reactions or because they are components of particular compounds, *e.g.* nitrogen in proteins, iron in haemoglobin, magnesium in chlorophyll, *etc.* Plants absorb most elements, other than carbon, oxygen and nitrogen in air, in selected chemical species from water, sediments and soil. Appropriate chemical species of these elements are absorbed by heterotrophs in their diet. This is an important consideration in assessing bioavailability.

Some elements, *e.g.* nitrogen and phosphorus (principally as nitrates and phosphates, respectively), may occur in low concentrations in the environment compared with the amounts needed by living organisms. Thus, availability of nitrogen and phosphorus may limit plant growth and primary production. If amounts of nitrate and phosphate are increased by their use as fertilizers in intensive farming, plant growth increases. Runoff of an excess of these nutrients into natural waters causes algal blooms that may release toxins. When they die, the algae are degraded by micro-organisms, often aerobically with depletion of oxygen that leads to anoxic conditions that kill fish.

Many elements may be at low environmental concentrations of bioavailable chemical species. However, this is not a problem if they are needed biologically only in small amounts. In contrast, all elements that are essential for life may be toxic when bioavailable in relatively large quantities. Since bioavailability depends upon chemical speciation, analysis of total elemental concentration may give a false impression of both bioavailability and potential toxicity.

Environmental Gradients. Any habitat has its own set of environmental conditions to which an organism must be tolerant if it is to occur there. Different species have different tolerances to physical and chemical environmental factors (abiotic factors), *e.g.* temperature, rainfall or soil nutrient status. The range of abiotic factors tolerated along a gradient of such factors can be considered as the fundamental (theoretical) niche of the species (Figure 24). In practice, species usually occupy a narrower range of conditions than this – the realized niche. They do not occur at the extremities of the theoretical range because interactions with other organisms (biotic interactions) inhibit them. For example, a species will be best adapted to the environment near to the middle of its tolerance range. Towards the extremities of the tolerance range, the species will be under stress. It will not compete successfully there with better-adapted species, which are surviving near the middle of their tolerance ranges.

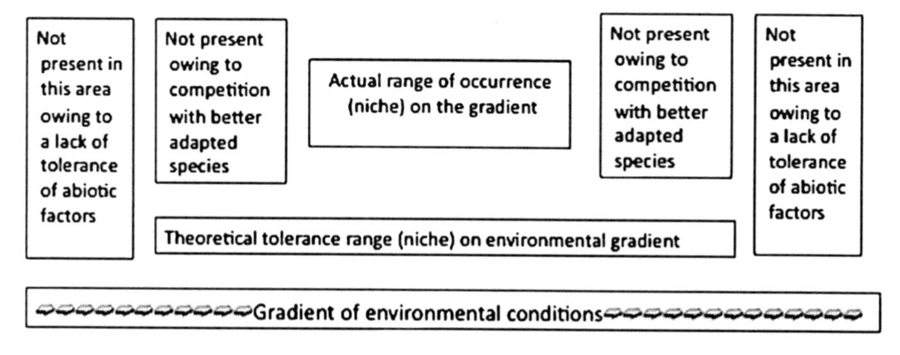

Figure 24 Illustration of a theoretical niche defined by an environmental gradient.

The Ecosystem Concept. An ecosystem consists of all the organisms in a particular place or habitat, their interrelationships with each other in terms of nutrient, carbon and energy flows, and in terms of biotic determinants of community composition, such as competition between species, the physical habitat and the abiotic factors associated with it. These factors also play a role in determining community composition and in determining primary and, hence, secondary production. Ecosystems can be quantified in terms of the fluxes of carbon, energy and nutrients, and the productivities of each trophic level. These properties can be quantitatively modelled using computers to enable predictions to be made about ecosystem performance.

The most important property of ecosystems is their 'dynamic stability' – their capacity to remain broadly the same over time in species composition and abundance, and in the magnitudes of processes, despite environmental variations. Although the climate fluctuates from year to year, the structure of an ecosystem tends to be stable within limits, and should be sustainable if the climatic fluctuations continue within their established limits. Any long period of global warming or cooling will cause continuing change until a new stable climatic situation is attained. One characteristic of dynamic stability is the maintenance over time of mean population sizes. In many insects reproduction occurs every year and the life span is one year or less. There can be fluctuations of several orders of magnitude in population size over several years but they fluctuate around a mean value. This may result from density-dependent factors, *i.e.* environmental factors whose intensity or effect depends on the population density. For example, at high density food may run short giving a population crash, while at low density the abundance of food may allow population size to increase, thus fluctuating about a mean over a period of years.

Ecosystem stability is not rigid. Systems change naturally. For example, on a short time scale, winter and summer aspects of a community in a temperate climate are very different. On a longer time scale there is ecological succession where one community naturally replaces another on an area of land or water, usually as a result of the modification of the habitat conditions by the organisms that are replaced so that the habitat is no longer suitable for their own survival. Thus, survivability is the basis of natural selection of species appropriate to the habitat conditions and hence of the evolution of species. This is generally explained on the basis of what has been called r/k selection theory. The terms r and k are derived from the Verhulst equation of population dynamics:

$$\frac{\mathrm{d}N}{\mathrm{d}t} = rN\left(1 - \frac{N}{k}\right)$$

where r is the growth rate of the population (N), and k is the carrying capacity of its local environmental setting. Typically, r-selected species exploit empty niches, and produce many offspring, each of which has a relatively low probability of surviving to adulthood. Organisms with r-selected traits range from bacteria and diatoms, through insects and weeds, to various cephalopods and mammals, especially small rodents. In contrast, *k*-selected species are strong competitors in crowded niches, and invest more heavily in fewer offspring, each of which has a relatively high probability of surviving to adulthood. Organisms with *k*-selected traits include large organisms such as elephants, humans and whales, but smaller organisms, such as Arctic terns, may also use this 'strategy' successfully. In the scientific literature, *r*-selected species are occasionally referred to as 'opportunistic', while *k*-selected species may be described as 'climax' or 'equilibrium' species.

From the above considerations, it may be seen that mature stable ecosystems must be characterized by a preponderance of *k*-strategists – the species that succeed by having a very precise adaptation to their environment. Earlier stages in a succession may have a greater proportion of *r*-strategists, the opportunists – organisms with wide environmental tolerance that do not survive so well in stable habitats when competing with more precisely adapted *k*-species. In stressful environments, caused either by human intervention or by naturally harsh conditions, tolerance to abiotic factors becomes a greater determinant of community composition than biotic interactions, and *r*-strategists predominate.

The above description of the concept of an ecosystem stresses the ability of such systems to remain stable within limits in various ways. Loss of this stability may be the most serious effect of potential toxicants on ecosystems at risk. Loss of stability may be associated with a temporary increase in species numbers and diversity. Thus, the common assumption that high species numbers and diversity is characteristic of a healthy ecosystem may not always be true. Management of ecosystems to maintain species balance may be more important than seeking to increase biodiversity.

Objective of Ecotoxicology. The objective of ecotoxicology is to define the concentration of chemicals at which organisms in the environment will be affected and to suggest how this concentration can be avoided in the environment by appropriate management. To study the possibility that a chemical is toxic, ecotoxicologists usually start with single species tests and progress to more tests on higher ecological levels. This process is referred to as 'tiered testing'. As testing takes in more of the complexity of ecological relationships, the resulting data become more relevant. However, it takes much more time and resources to get these data and thus only those substances used in very large amounts are tested at the highest tier, chronic testing in a model ecosystem.

For a population to flourish in its environment, individuals must survive to a size and age sufficient to permit reproduction. Effects of exposure to a potential environmental contaminant on survival, growth and reproduction are therefore the main concern of ecotoxicologists. In general, for organics, structure–activity relationship information will be considered before proceeding to tiered toxicity testing. Subsequently, the resultant toxicity data will be combined with environmental fate data to provide a basis for deciding what, if any, regulation and management is required. This process is called environmental risk assessment.

Environmental Risk Assessment. The first step in environmental risk assessment is the calculation of the predicted environmental concentration (PEC). In calculating this, both short-term exposures, such as accidental spillage, which may result in high concentrations in the environment for a relatively short time, and long-term exposures that are the result of continuing discharges, resulting in continuous and perhaps increasing environmental exposure, must be considered. Exposure assessment requires knowledge of the biodegradation profile of the substances entering the environment and the extent to which they may be removed by wastewater treatment. Taking these factors into account, the environmental distribution of organic compounds may be predicted by the application of mathematical models such as the fugacity models originally developed by Donald Mackay. These models have already been described in discussing 'persistence' (Section 4.1) and may be downloaded freely from the Trent University web site (*see* Further Reading).

The second step in environmental risk assessment is calculation of the predicted-no-effect-concentration (PNEC). This is the concentration that is believed to cause no adverse effect to organisms in the environment at risk. This concentration is calculated from toxicity tests on species relevant to each environmental compartment. For the aqueous environment, indicator species are typically a freshwater fish, a freshwater invertebrate and freshwater green algae. For sediments and soils, they are sediment dwelling organisms, earthworms and terrestrial plants. For air, the indicator species might be birds or insects of importance such as bees.

The PEC : PNEC ratio is used as an indicator of risk and is called the risk quotient (RQ) or risk index (RI). A substance is judged to be environmentally acceptable if the PNEC is higher than the PEC, *i.e.* if this value is less than 1 (PNEC > PEC). A simple environmental risk assessment may be completed in a few weeks or months, but comprehensive assessments for substances produced in large amounts may take years to complete. Even after an environmental risk assessment is completed, especially for high production volume (HPV) chemicals, it may be decided to monitor the concentrations of substances of concern to confirm the accuracy of the PNECs and PECs.

Further Reading

P. Bjerregaard and O. Andersen, Ecotoxicology of metals – sources, transport, and effects in the ecosystem, in *Handbook on the Toxicology of Metals*, ed. G. F. Nordberg, B.A. Fowler, M. Nordberg and L. T. Friberg, Elsevier, Amsterdam, 3rd edn, 2007, pp 251–280.

D. Mackay, *Multimedia Environmental Models, The Fugacity Approach*, Lewis Publishers, Boca Raton, FL, 1991. Available at <http://www.trentu.ca/academic/aminss/envmodel/models/L1L2L3.html>.

M. C. Newman and M. A. Unger, *Fundamentals of Ecotoxicology*, Lewis, Boca Raton, FL, 2nd edn, 2003.

G. M. Rand, *Fundamentals of Aquatic Toxicology*, Taylor & Francis, Washington, D.C., 2nd edn, 1995.

S. Sarkar, Ecology, in *The Stanford Encyclopedia of Philosophy*, ed. E. N. Zalta, Fall 2007 edition. Available at <http://plato.stanford.edu/archives/fall2007/entries/ecology/>.

Soil Biology Primer, online, July 19, 2008. Available at <http://soils.usda.gov/sqi/concepts/soil_biology/biology.html>.

P. F. Verhulst, *Corresp. Math. Phys.* 1838, **10**, 113.

C. H. Walker, S. P. Hopkin and R. M. Sibly, *Principles of Ecotoxicology*, CRC Press, Boca Raton, FL, 3rd edn, 2005.

Abbreviations, Acronyms and Initialisms

ADME	Absorption, distribution, metabolism and excretion
AIC	Akaike information criterion: a statistical procedure that provides a measure of the goodness-of-fit of a dose–response model to a set of data. The AIC is calculated from the equation $AIC = 2k - 2\ln(L)$, where k is the number of parameters and L is the likelihood function. Usually, normally distributed errors are assumed and the AIC is computed as $AIC = 2k + n\ln(RSS/n)$, where n is the number of observations and RSS is the residual sum of squares.
ANSI	American National Standards Institute
ATP	Adenosine triphosphate
BMC	Benchmark concentration
BMCL	Lower confidence limit for BMC
BMD	Benchmark dose
BMDL	Lower confidence limit for BMD
BMDS	Benchmark dose at a given standard deviation
BMR	Benchmark response
DNA	Deoxyribonucleic acid
EEA	European Environmental Agency
EC	Effective concentration
ECHA	European Chemicals Agency
ECB	European Chemicals Bureau
EDx	Effective dose for a biological effect in x% of the individuals in the test population
EFSA	European Food Safety Authority
GFR	Glomerular filtration rate
GHS	Globally Harmonized System of Classification and Labelling of Chemicals
HQ	Hazard quotient
IAEA	International Atomic Energy Agency
IPCS	International Programme on Chemical Safety
IUPAC	International Union of Pure and Applied Chemistry
JECFA	Joint FAO/WHO Expert Committee on Food Additives

LC_{50}	Median concentration lethal to 50% of a test population
LD_{50}	Median dose lethal to 50% of a test population
LEDx	Lowest effective dose for a biological effect in x% of the individuals in a test population
LOAEL	Lowest observed adverse effect level
MFO	Mixed-function oxidase
NADPH	Nicotinamide adenine dinucleotide phosphate (reduced)
NAG	*N*-Acetyl-D-glycosaminidase
NAS	National Academy of Science
NIOSH	U.S. National Institute of Occupational Safety and Health
NOAEL	No observed adverse effect level
PBPK	Physiologically-based pharmacokinetic modelling
PBPD	Physiologically-based pharmacodynamic modelling
PBTK	Physiologically-based toxicokinetic modelling
PEL	Permissible exposure limit
PIPs	Persistent inorganic pollutants
PK	Pharmacokinetic
POP	Persistent Organic Pollutant
QSAR	Quantitative structure–activity relationship
QSMR	Quantitative structure–metabolism relationship
REACH	Registration, Evaluation, Authorisation, and Restriction of Chemicals
RfC	Reference concentration
RfD	Reference dose
ROS	Reactive oxygen species
SAR	Structure–activity relationship, specific (standard) absorption rate
SD	Standard deviation
SMR	Structure–metabolism relationship
SE	Standard error
TEF	Toxicity equivalency factor
TEQ	Toxicity equivalent
UDP	Uridine diphosphate
UNECE	United Nations Economic Commission for Europe
USEPA	United States Environment Protection Agency
USFDA	United States Food and Drug Agency
WR	Quality factor (radiation)

Subject Index

Breinigsville, PA USA
23 May 2010
238520BV00005B/3/P

9 780854 041572